2

Focus on Phonics

SHORT VOWEL SOUNDS AND CONSONANT BLENDS

BY GAIL V. RICE

ProLiteracy
New Readers Press®

Focus on Phonics 2
ISBN 978-1-56420-943-6

Published by New Readers Press, a division of ProLiteracy.

New Readers Press
308 Maltbie Street, Suite 100, Syracuse, NY 13204
newreaderspress.com | proliteracy.org

Printed in the United States of America
20 19 18 17 16 15 14

Proceeds from the sale of New Readers Press materials support professional development, training, and technical assistance programs of ProLiteracy that benefit local literacy programs in the U.S. and around the globe.

Developmental Editor: Terrie Lipke
Creative Director: Andrea Woodbury
Production Specialist: Maryellen Casey
Art and Design Supervisor: James P. Wallace
Illustrations: Bill Petersen, Ron Mahoney, and Drew Rose, represented by Wilkinson Studios, Inc.
Cover Design: Carolyn Wallace

Table of Contents

Part A: Short Vowel Sounds

Part B: Consonant Sounds

Do They Rhyme?

Look at the pictures and say the words.
Do the words rhyme?
Make a check ☑ in the box if they do.

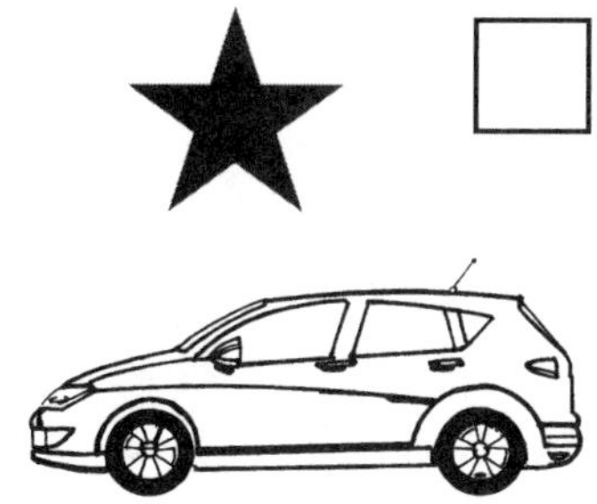

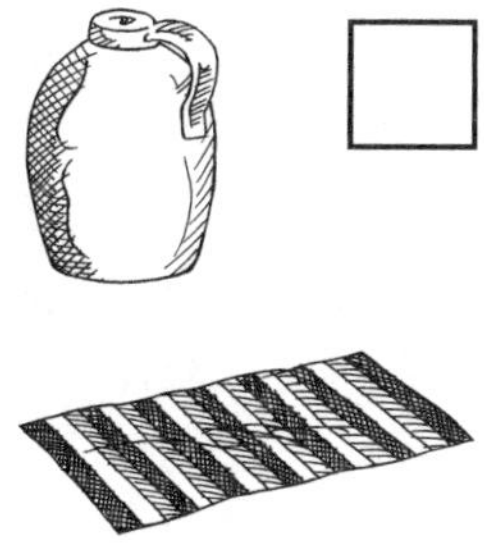

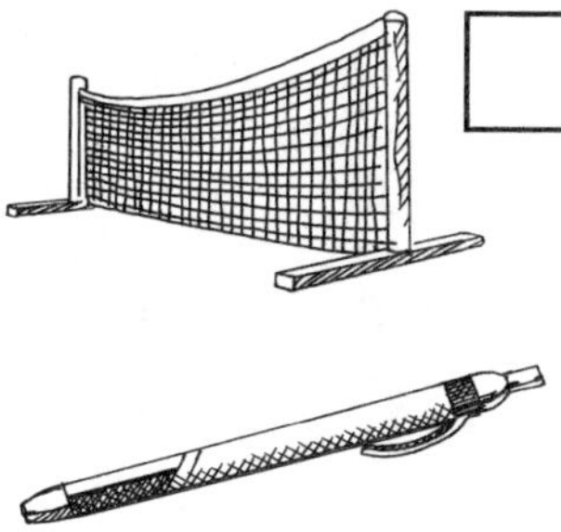

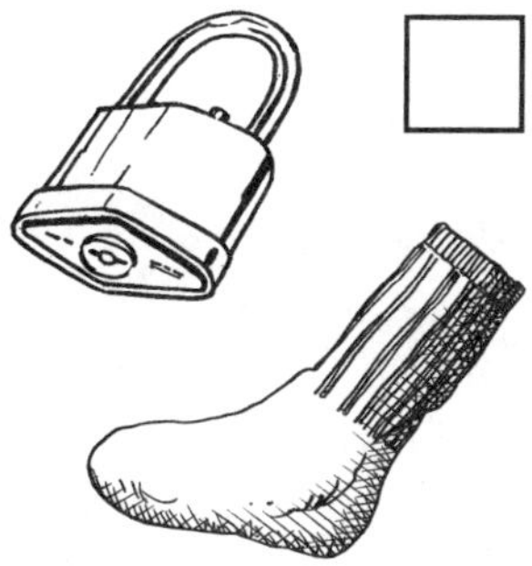

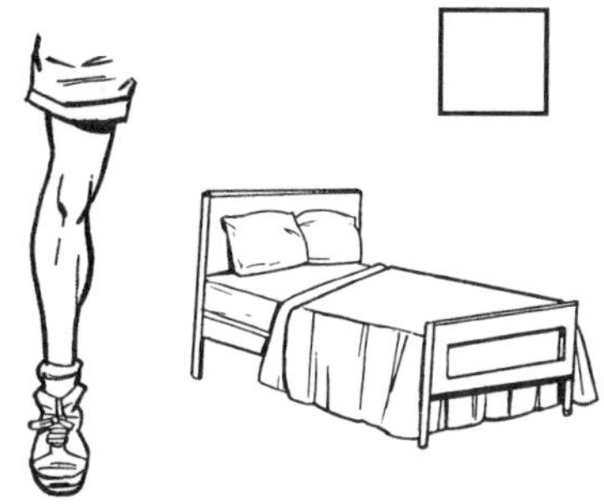

Two Forms of the Letter **a**

1. | a = a |
|---|

at	at
and	and
man	man
has	has

2. Circle all the words that are spelled the same as the first one.

an	an	on	an	in	and	an
are	and	are	an	are	at	are
at	at	in	at	is	on	at
has	his	has	hand	has	has	her
pan	an	pan	pat	hand	pan	Pam

3. Write these words.

an	an
are	
pan	
hand	
apple	
thank	

4. Read these sentences.

1. The man has a pan in his hand.
2. Cal looks at the girl's hands.
3. Dan and Ann are at Pam's shop.
4. Cal has a quarter for Dan.
5. Sam gives an apple to Van.
   Van thanks Sam.

Practice 4-A: Word Families -un, -um

1.

run	gum
gun	**-um**
sun	
-un	

2. Write the letter or letters and say the word.

b ____ un	b ____ um
f ____ un	g ____ um
g ____ un	h ____ um
n ____ un	r ____ um
r ____ un	s ____ um
s ____ un	

3. Read the words.

nun	gum
sun	rum
gun	bum
run	hum
bun	sum
fun	

4. Write the word you hear.

1. ____________	7. ____________
2. ____________	8. ____________
3. ____________	9. ____________
4. ____________	10. ____________
5. ____________	11. ____________
6. ____________	

5. Review Words

son hunt

New Word

out

Words from skill book 1

Ann children do sell

6. Read the sentences.

1. Kim will sing, and I will hum.
2. He puts the fish on a bun.
3. My son thinks it is fun to hunt.
   He is out hunting with his gun.
4. The man is a bum.
5. Do they sell gum in this shop?
6. Sister Ann is a nun.
7. The sum of 2 and 4 is 6.
8. Run and get my gun.
9. He has a drink of rum.
10. Is the sun up?
11. The nun hums and sings.
12. The children have fun out in the sun.
13. Do not run with a gun in your hand.
14. Sum it up for me.
15. My son puts the buns in the dish.
16. He wants me to give him gum.
    I do not have gum. I have run out of it.

Practice 4-B: Word Families -up, -ub, -us, -uss

1.

cup	rub	bus	fuss
pup	**-ub**	**-us**	**-uss**
-up			

2. Write the letter or letters and say the word.

	up		us
c	____ up	b	____ us
p	____ up	th	____ us
c	____ ub	c	____ uss
r	____ ub	f	____ uss
t	____ ub		

3. Read the words.

up	rub	bus	fuss
cup	tub	us	cuss
pup	cub	thus	

4. Write the word you hear.

1. ____________
2. ____________
3. ____________
4. ____________
5. ____________
6. ____________
7. ____________
8. ____________
9. ____________
10. ____________
11. ____________

5. Review Words

son hunt

New Words

husband just

Words from skill book 1

go jump tell under woman

6. Read the sentences.

1. The woman fills up the tub.
2. He rubs the pup's neck.
3. My husband is picking up our pup.
4. Put the cups in the sink.
5. Just six of us will go on this bus.
   He will just miss his bus.
6. The woman rubs her hands.
7. Do not cuss at us!
8. The little cubs jump and run.
9. The kids will fuss if they do not go with us.
10. The pups run under the bus.
11. My husband is sick. Thus, he will not go with us.
12. The woman has just two cups.
13. My husband runs up the street to get the bus.
14. The woman sits in the tub.
15. If your leg hurts, rub it.
16. The woman tells her husband, "Do not fuss at me."
17. Did the Cubs win?

Practice 4-C: Word Families -ut, -ud, -uff

1. cut mud puff

-ut **-ud** **-uff**

2. Write the letter or letters and say the word.

b ___ ut	b ___ ud	
c ___ ut	m ___ ud	
h ___ ut	B ___ ud	
r ___ ut		
n ___ ut	c ___ uff	
sh ___ ut	m ___ uff	
	p ___ uff	

3. Read the words.

hut	Bud	puff
rut	mud	muff
nut	bud	cuff
but		
shut		
cut		

4. Write the word you hear.

1. ____________	7. ____________
2. ____________	8. ____________
3. ____________	9. ____________
4. ____________	10. ____________
5. ____________	11. ____________
6. ____________	12. ____________

5. Review Words

duck hunt Jimmy

Words from skill book 1

box no

6. Read the sentences.

1. Cut up the duck for us.
2. Jimmy gets mud on his cuff.
3. She shut up the shop.
4. Kitty is looking at the little pink buds.
   The buds are little, but they will get big.
5. Bill has no cuts on his hands.
6. The man lives in a hut.
   The hut has mud on it.
7. The woman gives nuts and gum to the kids.
8. I want to hunt ducks, but I have no gun.
9. Bud will not shut the box.
10. She puts her hands into the muff.
11. He gets a cut on his lip.
    His lip will puff up and get big.
12. The bus is in a big rut, but we will get it out.
13. We had fun hunting for nuts.
14. The woman tells the kids to shut up.
15. No one but Jill gets cut.

Practice 4-D: Word Families -ug, -udge

1. rug judge
jug **-udge**
-ug

2. Write the letter or letters and say the word.

b ____ ug
d ____ ug
h ____ ug
j ____ ug
m ____ ug
r ____ ug
t ____ ug

b ____ udge
f ____ udge
j ____ udge
n ____ udge

3. Read the words.

hug
tug
mug
bug
jug
dug
rug

fudge
nudge
budge
judge

4. Write the word you hear.

1. ____________
2. ____________
3. ____________
4. ____________
5. ____________
6. ____________
7. ____________
8. ____________
9. ____________
10. ____________
11. ____________

5. New Word

water

6. Read the sentences.

1. The woman hugs her son.
2. The jug has rum in it.
3. Is he a judge?
4. He tugs on the big rug.
5. The pup dug in the mud.
6. Put the water in a mug, not in a cup.
   Fill the mug with water.
7. I give Bud a nudge.
8. A big bug is under the rug.
9. Bill gives Kim a box of fudge.
   The fudge has nuts in it.
10. I hug and kiss my husband.
11. He will judge the pigs.
12. My son tugs on my hand.
13. I tug on the box, but it will not budge.
14. This cup of water has a bug in it.
15. Her pup gets mud on the rug.
16. He drinks out of the water jug.

Practice 5-A: Word Families -uck, -ush, -uch

1.

duck	rush	much
Buck	**-ush**	**-uch**
-uck		

2. Write the letter or letters and say the word.

b ____ uck	g ____ ush
d ____ uck	h ____ ush
l ____ uck	r ____ ush
s ____ uck	m ____ ush
t ____ uck	
st ____ uck	m ____ uch
tr ____ uck	s ____ uch
B ____ uck	
Ch ____ uck	

3. Read the words.

tuck	duck	mush	such
luck	suck	hush	much
buck	truck	gush	
stuck	Buck	rush	
Chuck			

4. Write the word you hear.

1. ______________	9. ______________
2. ______________	10. ______________
3. ______________	11. ______________
4. ______________	12. ______________
5. ______________	13. ______________
6. ______________	14. ______________
7. ______________	15. ______________
8. ______________	

5. Review Words

does from come some

New Words

so good

Words from skill book 1

thank visit

6. Read the sentences.

1. Bud Buck comes from the city in his truck.
   Does his truck get stuck in the mud?
2. Kitty is fixing me some ducks for dinner.
   I tell her, "Thanks so much for such a good dinner."
3. Tuck it in.
4. Bud has such good luck.
5. He sells us some apples for two bucks.
6. Do not rush so much!
7. It is so good to have Chuck come to visit us.
8. Sam and Jim rush out of the shop.
9. Do not put so much mush in my dish.
   I just want a little.
10. The gum gets stuck on my fingers.
11. Mrs. Buck tells her children to hush.
12. He does not have good luck fixing trucks.
13. The fudge is stuck to the pan.
14. The little girl sucks on her finger.
15. Chuck has good luck hunting. He gets a big buck.

Practice 5-B: Word Families -ung, -unk

1.

sung	sunk
-ung	junk
	-unk

2. Write the letter or letters and say the word.

h ____ ung	b ____ unk
l ____ ung	d ____ unk
r ____ ung	h ____ unk
s ____ ung	j ____ unk
	p ____ unk
	s ____ unk
	ch ____ unk

3. Read the words.

rung	junk	chunk
hung	sunk	dunk
sung	hunk	punk
lung	bunk	

4. Write the word you hear.

1. ____________	7. ____________
2. ____________	8. ____________
3. ____________	9. ____________
4. ____________	10. ____________
5. ____________	11. ____________
6. ____________	

5. Review Words

funny	mother	brother
uncle	picture	word

New Words

other like

6. Read the sentences.

1. My brother likes to sit on his bunk.
2. Uncle Bud will get rid of the junk.
3. Dunk it under the water.
4. Mother hung up one other picture.
5. My brother Chuck has sung for us.
   We like the funny words he has sung.
6. A fish has no lungs.
7. Bill likes to dunk his sister in the water.
   The little punk thinks it is funny!
8. This hunk of tin is no good to us.
   Put it with the other junk.
9. The ship has sunk under the water.
10. Mother cuts a big chunk of fudge.
11. The picture from my uncle looks like junk to us.
    It looks funny, so we have not hung it up.
12. The telephone has rung. Did she get it?
    She did pick it up, but they hung up.
13. I want this bunk, not the other one.

Practice 5-C: Adding Endings

1. Adding *-ing* to VCC words

V C C + i n g

h u n t + i n g ⟶ hunting

f i l l + i n g ⟶ filling

th i n k + i n g ⟶ thinking

2. Label the last three letters. Write the word with *-ing.*

jump ______________________

kiss ______________________

fish ______________________

bring ______________________

3. Add *-ing* to the word. Read the sentence.

drink 1. Are you ____________ this?

hurt 2. Bud's leg is ____________ him.

dunk 3. He is ____________ it in water.

4. Adding *-ing* to CVC words

C V C + C + i n g

r u n + n + i n g ⟶ running

s hi p + p + i n g ⟶ shipping

h u m + m + i n g ⟶ humming

5. Label the last three letters. Write the word with *-ing.*

hug ______________________

bid ______________________

dig ______________________

dip ______________________

6. Add *-ing* to the word. Read the sentence.

hug 1. Mother is ____________ her kids.

shut 2. She is ____________ up the shop.

run 3. Dick is ____________ up the hill.

7. Label the last three letters with VCC or CVC. Write the word with *-ing.* Read the word.

rush __________	chill __________
bud __________	cut __________
get __________	fill __________
pick __________	rub __________
lick __________	rush __________
want __________	tip __________

8. Take off the *-ing.* Write the root word. For some words you may have to take off a letter.

ripping ______________________

pinning ______________________

shutting ______________________

fussing ______________________

wishing ______________________

Practice 5-D: Contrasting Short Vowels i and u

i	**rib**	**dig**	**lick**	**bid**	**him**	**sing**
u	**rub**	**dug**	**luck**	**bud**	**hum**	**sung**

1. Write the letter.

1. fin
   fun
   f ____ n

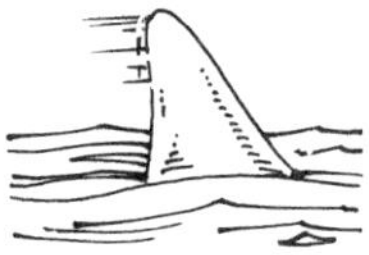

3. sin
   sun
   s ____ n

5. sink
   sunk
   s ____ nk

2. rig
   rug
   r ____ g

4. hit
   hut
   h ____ t

2. Circle the word that you hear.

1. big bug	4. miss muss	7. bit but	10. hitch hutch	13. this thus
2. lick luck	5. bin bun	8. chick chuck	11. dig dug	14. pin pun
3. rib rub	6. him hum	9. rim rum	12. sick suck	15. ring rung

3. Write the short vowel that you hear.

1. t ____ ck	4. m ____ ff	7. s ____ ng	10. f ____ zz	13. c ____ d
2. s ____ p	5. p ____ g	8. j ____ g	11. p ____ t	14. ch ____ n
3. th ____ nk	6. b ____ d	9. h ____ nk	12. r ____ t	15. s ____ ch

4. Circle the right word.

1. She will rib / rub her hands.
2. Bring me the pink / punk dish.
3. The pup bit / but my leg.
4. We wish him good lick / luck .
5. I sing and him / hum to the children.
6. The ring / rung is on her finger.

Practice 5-E: Review Lesson – Short u

1. Look at the picture and say the word. Then fill in the right letter to make the word.

h
___ ug
r

n
bu ___
m

ff
pu ___
p

t
cu ___
b

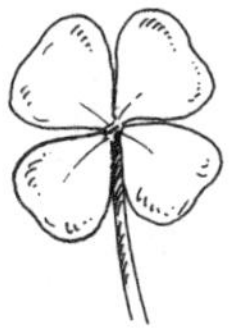

l
___ uck
p

f
___ udge
j

2. Circle the word that makes sense in the sentence.

1. A bug bit his hand. Did his hand puck / puff up?
2. Do not rub / cub the mud on the rug.
3. Bud put the box under his buck / bunk .
4. The hum / gum is stuck on my hands.
5. Has she sung / sunk for us?
6. The bum is drinking some rum / run .

3. Look at the picture and say the word. Then circle the right word.

1.  mug hug jug rug
2. 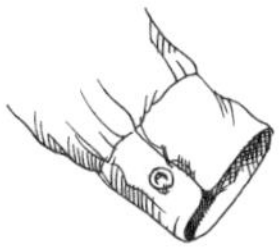 puff cut cuff cuss
3.  sung sun sum sunk
4.  buzz bus buck but
5.  cub up cut cup
6.  nut run nun fun

Practice 5-E: Review Lesson – Short u (continued)

4. Look at the picture and say the word. Then write the word under the picture.

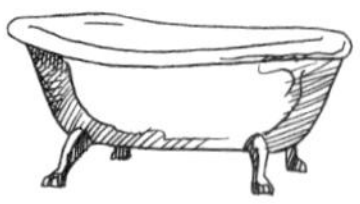 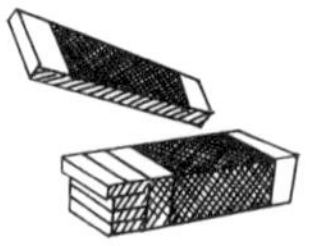 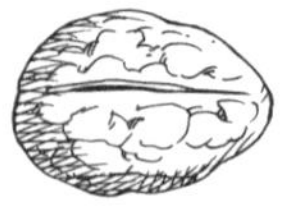

________ ________ ________

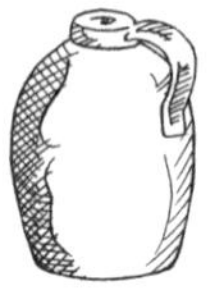  

________ ________ ________

5. Circle all the words that are the same as the first. Work from left to right.

tug	lug	tug	tub	gut	tug	hug	tug
but	bud	but	dub	bud	dud	bid	bug
cut	tuck	out	cut	cub	rut	put	cut
gum	gun	hum	gum	bum	mug	gum	gun
luck	luck	lung	lick	tuck	luck	look	buck
run	sun	nun	run	rum	run	rung	run
bus	bun	bus	sub	pus	bus	us	sub
pup	pub	pup	cup	dud	pug	pup	puff

6. Read these sentences.

1. Mrs. Buck puts one cup of nuts in the fudge.
2. Chuck's bus is stuck in a big rut.

   But my uncle will get it out with the truck.
3. The kids fuss until Mother says, "Hush!"
4. I hung up a picture of my uncle. He is a judge.
5. This rug is not much good. Put it out with the junk.
6. We had such fun hunting and fishing.
7. I rush up to my son and hug him.

Practice 6-A: Word Family -ell

1. yell
sell
well
-ell

2. Write the letter or letters and say the word.

b ____ ell	w ____ ell
f ____ ell	y ____ ell
h ____ ell	sh ____ ell
s ____ ell	B ____ ell
t ____ ell	N ____ ell

3. Read the words.

well	Bell
bell	shell
sell	Nell
hell	fell
tell	yell

4. Write the word you hear.

1. ____________	6. ____________
2. ____________	7. ____________
3. ____________	8. ____________
4. ____________	9. ____________
5. ____________	10. ____________

5. Review Words

help twelve very fresh many

New Words

until jelly

Words from skill book 1

egg seven

6. Read the sentences.

1. The woman tells us she has twelve children.
2. Mr. Bell has many fresh eggs to sell.
   He sells us a box of twelve fresh eggs.
3. I will help the sick man until he gets well.
4. Bud yells to Mr. Bell, "Help! Help!"
5. She fell and hurt her leg.
6. We will not have dinner until seven.
7. Nell picks up many shells from the water.
8. The woman sings very well.
9. The bell will ring at twelve.
10. He does not want to go to hell.
11. She yells, but no one comes to help.
12. Get the nut out of the shell.
13. Nell sells very good jelly at her shop.
    Bud tells us to get some apple jelly from her.
14. The little girl fell into a big well.
15. Tell your brother not to yell at me.
16. The children yell until the bell rings.

Practice 6-B: Word Family -et

1. get

   pet

   -et

2. Write the letter or letters and say the word.

b ____ et	p ____ et
g ____ et	s ____ et
j ____ et	v ____ et
l ____ et	w ____ et
m ____ et	y ____ et
n ____ et	

3. Read the words.

set	get	bet
pet	vet	met
net	let	jet
yet	wet	

4. Write the word you hear.

1. ____________	7. ____________
2. ____________	8. ____________
3. ____________	9. ____________
4. ____________	10. ____________
5. ____________	11. ____________
6. ____________	

5. **Review Words**

   cents seventy

   New Words

   any must there TV

   Words from skill book 1

   says yes

6. Read the sentences.

   1. Jimmy says a pup is a very good pet.
   2. Mr. Bell will get on a big jet.
   3. Set the eggs there.
   4. Have you met my brother yet?
   5. Mother will not let the kids look at TV.
   6. Are there any fish in the net?
   7. He must rush his pet to the vet.
      The vet will set the pup's leg.
   8. Bill will bet seventy cents.
   9. Yes, he met her there for dinner.
   10. I must sell my TV set.
   11. The pup jumps into the tub and gets wet.
   12. He does not let the children have any pets.
   13. You must fix the net if you want to get any fish.
   14. I bet he will not get there until twelve.
   15. The sun has not set yet.
   16. Yes, let us go there on a jet.
   17. I do not want to bet.

Practice 6-C: Word Families -en, -eck

1.

hen	neck
ten	**-eck**
men	
-en	

2. Write the letter or letters and say the word.

d ____ en	th ____ en	d ____ eck
h ____ en	wh ____ en	n ____ eck
m ____ en	B ____ en	p ____ eck
p ____ en	K ____ en	ch ____ eck
t ____ en		

3. Read the words.

men	pen	peck
ten	then	neck
Ben	Ken	check
hen	when	deck
den		

4. Write the word you hear.

1. __________	8. __________
2. __________	9. __________
3. __________	10. __________
4. __________	11. __________
5. __________	12. __________
6. __________	13. __________
7. __________	

5. Review Words

cents twelve

New Word

women

Words from skill book 1

four Glenn Pam tent write

6. Read the sentences.

1. The deck of the ship is wet.
2. Pam is a woman. Pam and Ann are women.

   Ken is a man. Ken and Ben are men.
3. Glenn writes a check for the TV set.
4. Ken is sitting in his den with two men.
5. This pen is just ten cents, but it writes well.
6. If the bus does not run, then the men will check it.
7. The chick pecks at the egg to get out of it.

   The little chick will get to be a big hen.
8. When Ben is four, then Ken will be ten.
9. When will the dinner be? At seven.

   Twelve men and ten women are coming then.
10. Ben's neck hurts.
11. The little cubs live in the den.
12. The women write checks when they shop.
13. Ten big hens are in the pen.
14. Does the tent have a rip? Check it.
15. Men and women are sitting on the deck.

Practice 7-A: Word Families -eg, -ess

1. leg less

-eg **-ess**

2. Write the letter or letters and say the word.

		egg	gu	____	ess
b	____	eg	l	____	ess
k	____	eg	m	____	ess
l	____	eg	ch	____	ess
p	____	eg	B	____	ess
M	____	eg	J	____	ess
P	____	eg			

3. Read the words.

peg	beg	guess
leg	Meg	less
egg		Jess
keg	mess	chess
Peg	Bess	

4. Write the word you hear.

1. ____________	8. ____________
2. ____________	9. ____________
3. ____________	10. ____________
4. ____________	11. ____________
5. ____________	12. ____________
6. ____________	13. ____________
7. ____________	

5. Review Words

quickly pretty them

New Words

best dress make unless

Words from skill book 1

address nest valley

6. Read the sentences.

1. Meg will get the best dress at the shop.
   The dress is selling for much less.
2. Peg gives a chess set to them.
3. He will run unless his leg hurts.
4. You must have the pegs to put up the tent.
5. Jess is bringing a keg of rum.
6. I will have to guess unless you tell me.
7. Bess begs the kids not to make a mess.
8. Yes, I guess this box of eggs is the best.
9. Write your address on the check.
10. It is less windy in the valley.
11. Mother tells them to pick up the mess quickly.
12. Jess will get some eggs from the hens' nests.
13. Did Ken make the best guess?
14. I will not win at chess unless you help me.
15. Your dress makes you look very pretty.
16. The man begs them for a quarter.
17. Peg makes a mess in the kitchen.

Practice 7-B: Word Families -ed, -edge

1.

Ed	edge
bed	ledge
red	**-edge**
-ed	

2. Write the letter or letters and say the word.

b ____ ed	Ed	edge
f ____ ed	Fr ____ ed	h ____ edge
l ____ ed	N ____ ed	l ____ edge
r ____ ed	T ____ ed	w ____ edge
w ____ ed		
sh ____ ed		

3. Read the words.

led	red	hedge
wed	Ted	edge
shed	Ned	wedge
Ed	fed	ledge
bed	Fred	

4. Write the word you hear.

1. ____________	8. ____________
2. ____________	9. ____________
3. ____________	10. ____________
4. ____________	11. ____________
5. ____________	12. ____________
6. ____________	13. ____________
7. ____________	14. ____________

5. Review Words

send friend letter

New Words

sent went wedding

6. Read the sentences.

1. Did Ed send letters to any of his friends?

   Yes, he sent a letter to his friend Ted.
2. Ned went out to the shed to get a pick.
3. Fred will cut the hedge for me.
4. Mother fed her children and sent them to bed.

   They went to bed at ten.
5. Meg wants a pretty wedding.
6. Ted went to the city in his red truck.
7. The bird makes a nest on the ledge.
8. Ned and Ted fed the pets.
9. He led them to the edge of the river.
10. Does your pup shed?
11. Bess sent Meg a pretty red dress.
12. When will Ned and Peg be wed?

    I want to go to the wedding and send a gift.
13. Fred sits on the edge of his bed.
14. Ed's neck gets red when he sits in the sun.
15. This wedge has a thin edge.

Practice 7-C: Adding Endings

1. To add *-ing* to a word that ends with *-e*, take off the *-e* and add *-ing*.

g i v ~~e~~ + i n g ⟶ giving

n u d g ~~e~~ + i n g ⟶ nudging

m a k ~~e~~ + i n g ⟶ making

2. Add *-ing* and read the word.

live ____________________

have ____________________

come ____________________

whistle ____________________

judge ____________________

like ____________________

3. Take off the *-e*, and add *-ing* to the word under the blank. Read the sentences.

1. Bess is ____________ letters.
   write
2. Ed is ____________ the pigs.
   judge
3. We are ____________ apple jelly.
   make

4. Reviewing endings
Take off the *-s* ending, and read the word.

whips ____________________

sets ____________________

makes ____________________

puffs ____________________

quits ____________________

says ____________________

brings ____________________

tells ____________________

ticks ____________________

5. Reviewing endings
Add *-ing* to the word and fill in the blank. (For some words you may have to take off the *-e*. For some words you will have to double the last letter.) Read the sentences.

1. send — They are __________ you a check.
2. bet — Fred is __________ seventy cents.
3. dress — Peg is __________ her little kids.
4. come — Dick is __________ to dinner.
5. whistle — Ted is __________ for the pup.
6. let — She is __________ us go to the city.
7. check — The men are __________ the truck.
8. rush — Jimmy is __________ to the shop.
9. take — The kids are __________ the cups to the sink.

6. Reviewing endings
Take off the *-ing* ending and write the root word. (For some words you may have to add a final *-e*. For some words you may have to take off a letter.)

making ____________________

shedding ____________________

writing ____________________

thinking ____________________

taking ____________________

hugging ____________________

bringing ____________________

dipping ____________________

coming ____________________

Practice 7-D: Contrasting Short Vowels e and i

i	**bill**	**lit**	**pick**	**him**	**big**	**miss**
e	**bell**	**let**	**peck**	**hem**	**beg**	**mess**

1. Write the letter.

1. pig
   peg
   p ____ g

3. bid
   bed
   b ____ d

5. pin
   pen
   p ____ n

2. will
   well
   w ____ ll

4. lid
   led
   l ____ d

2. Circle the word that you hear.

1. bit bet	4. big beg	7. rid red	10. pick peck	13. tin ten
2. till tell	5. wit wet	8. lit let	11. fill fell	14. pit pet
3. chick check	6. him hem	9. miss mess	12. six sex	15. bill bell

3. Write the short vowel that you hear.

1. y ____ t	4. w ____ ll	7. l ____ d	10. d ____ n	13. s ____ ll
2. h ____ p	5. p ____ g	8. th ____ ck	11. p ____ p	14. k ____ ss
3. d ____ ck	6. w ____ n	9. r ____ m	12. ____ tch	15. f ____ n

4. Circle the right word.

1. Did you get the (bill / bell) for the dress?
2. Jill wants (tin / ten) apples.
3. She will (pick / peck) the big ones.
4. I want to go with (him / hem) .
5. You will (miss / mess) us when we go.
6. (Lit / Let) me go with you.

Practice 7-E: Contrasting Short Vowels e and u

u	**bud**	**hum**	**lug**	**pup**	**duck**	**but**
e	**bed**	**hem**	**leg**	**pep**	**deck**	**bet**

1. Write the letter.

1. chuck
   check
   ch ____ **ck**

3. pup
   pep
   p ____ **p**

5. duck
   deck
   d ____ **ck**

2. bug
   beg
   b ____ **g**

4. nut
   net
   n ____ **t**

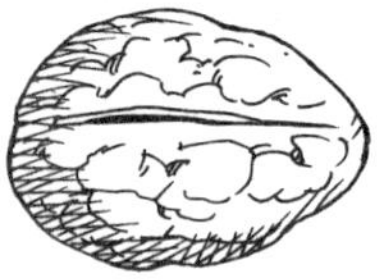

2. Circle the word that you hear.

1. hum hem	4. muss mess	7. duck deck	10. pun pen	13. mush mesh
2. bug beg	5. bud bed	8. lug leg	11. nut net	14. chuck check
3. jut jet	6. pup pep	9. but bet	12. dun den	15. bun Ben

3. Write the short vowel that you hear.

1. p ____ t	4. m ____ g	7. t ____ b	10. b ____ s	13. m ____ t
2. b ____ ck	5. l ____ ss	8. p ____ ff	11. f ____ d	14. p ____ g
3. k ____ g	6. b ____ t	9. wh ____ n	12. r ____ sh	15. g ____ m

4. Circle the right word.

1. He is sick, (but / bet) he will get well.
2. Put some (mush / mesh) in the dish.
3. (Bun / Ben) and Rick are fishing.
4. The little girl has a (pup / pep).
5. Put some (nuts / nets) in the fudge.
6. The man sits in his (dun / den).

Practice 7-F: Review Lesson – Short e

1. Look at the picture and say the word. Then fill in the right letter to make the word.

sh / ch ____ ess

ne ____ ck / t

g / j ____ et

d / p ____ eck

she ____ d / ll

t / d ____ en

2. Circle the word that makes sense in the sentence.

1. Ned has not met my sister wet / yet .
2. Fred yells, "This shed is a less / mess !"
3. Ed will cut the hedge for seventy cents / sent .
4. Jess fell / fed and hurt his leg.
5. The children beg / peg Mother for a pet.
6. The vet will check my peck / pet snake.

3. Look at the picture and say the word. Then circle the right word.

1.  peg beg leg keg
2.  yell yet shell well
3.  hem then hen when
4. 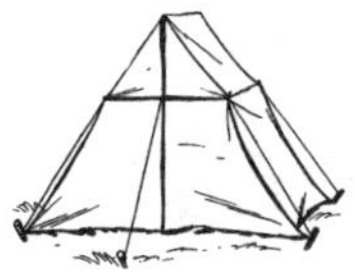 tell tent sent ten
5.  fed shell bed shed
6.  bet bed bell hell

Practice 7-F: Review Lesson – Short e (continued)

4. Look at the picture and say the word. Then write the word under the picture.

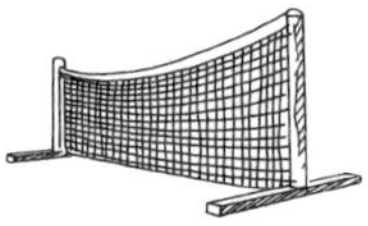

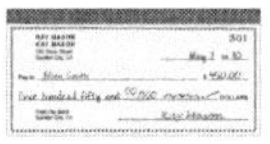

5. Circle all the words that are the same as the first. Work from left to right.

peck	peck	pick	beck	peck	deck	peck	pet
ten	hen	ten	tent	then	tell	ten	net
less	chess	sell	less	led	less	let	mess
bed	bed	bet	deb	bell	bed	bud	led
wet	met	vet	wet	well	wet	met	yet
leg	led	leg	beg	get	lug	keg	leg
sell	sell	set	less	sell	well	cell	sell
them	then	them	hem	then	met	the	them

6. Read these sentences.

1. Peg tells me, "Ted is sick, but he will get well."
2. Yes, the men sell fresh eggs there.
3. Ken led his friends to the edge of the river.
4. Meg fed her kids. Then the kids went to bed.
5. Ed sits in the den looking at the TV set.
6. This pen has red ink. It sells for twelve cents.
7. If you make the best guess, you win the bet.
8. When Bess makes jelly, she lets the kids help her.

Practice 8-A: Word Family -at

1. bat

cat

-at

2. Write the letter or letters and say the word.

	at	p ____	at
b ____	at	r ____	at
c ____	at	s ____	at
f ____	at	ch ____	at
h ____	at	th ____	at
m ____	at	P ____	at

3. Read the words.

hat	at	chat
mat	fat	rat
pat	that	cat
bat	Pat	sat

4. Write the word you hear.

1.	____________	7.	____________
2.	____________	8.	____________
3.	____________	9.	____________
4.	____________	10.	____________
5.	____________	11.	____________
6.	____________	12.	____________

5. New Words

shall take

Word from skill book 1

Cal

6. Read the sentences.

1. I will take this hat, not that one.
2. Pitch it, and I will hit it with the bat.
3. Bess and Pat like to chat on the telephone.
4. There is mud on that mat.
5. That cat is getting fat.
   Shall I take the cat to the vet?
6. Cal sat on my hat.
7. There are rats in that shed.
8. That dress will not fit if I get fat.
9. That bat has big wings.
10. Do not put fat in the sink.
11. Pat sells cats at her pet shop.
12. The pup sat on the mat.
13. Cal pats my hand.
14. Shall I take that bat to my brother?
15. She wants to chat with Cal.
16. Bud is fat, but he wants to be thin.
17. Shall we chat for a little bit?

Practice 8-B: Word Family -an

1. an
man
pan
-an

2. Write the letter or letters and say the word.

	an	t	____ an
b	____ an	v	____ an
c	____ an	th	____ an
f	____ an	D	____ an
m	____ an	J	____ an
p	____ an	N	____ an
r	____ an	V	____ an

3. Read the words.

fan	tan	pan
ran	can	Jan
man	Van	ban
van	an	than
Dan	Nan	

4. Write the word you hear.

1. ____________	8. ____________
2. ____________	9. ____________
3. ____________	10. ____________
4. ____________	11. ____________
5. ____________	12. ____________
6. ____________	13. ____________
7. ____________	14. ____________

5. Review Words

cannot black stand

New Words

as by

Word from skill book 1

Ann

6. Read the sentences.

1. Jan is standing by the fan.
2. Put an egg into the pan of water.
3. Jan has a big tan van.
4. Nan does as well as she can.
5. Ann can fix the kitchen fan.
6. Dan ran as well as the other man.
7. Jan cut up an apple and put it in the pan.
8. Dan gets as much sun as Pat.
   But Dan is less tan than Pat.
9. Did you yell to him as he ran by?
10. There is a ban on fishing. You cannot do it.
11. The red van ran into the tan truck.
12. This set of pans sells for less than that set.
13. Get a can of black olives.
14. The man is so sick that he cannot stand up.
15. Can you take the big fan there in the van?
16. I sit out in the sun and get a tan.
17. You can set the dish by the pans.

Practice 8-C: Word Families -am, -ab

1. Sam cab

-am **-ab**

2. Write the letter or letters and say the word.

		am	c	____	ab
d	____	am	d	____	ab
h	____	am	g	____	ab
j	____	am	j	____	ab
r	____	am	l	____	ab
y	____	am	n	____	ab
P	____	am	t	____	ab
S	____	am			

3. Read the words.

jam	dam	tab	nab
Pam	ram	jab	gab
am	yam	cab	lab
Sam	ham	dab	

4. Write the word you hear.

1. ____________ 9. ____________
2. ____________ 10. ____________
3. ____________ 11. ____________
4. ____________ 12. ____________
5. ____________ 13. ____________
6. ____________ 14. ____________
7. ____________ 15. ____________
8. ____________

5. Review Words

back family building

New Word

keep

6. Read the sentences.

1. Sam takes a cab and gets there quickly.
2. Pam is fixing this ham for her family.
   They will have yams with the ham.
3. Do you want some jam?
   Yes, I will have a little dab.
4. Help me! I am in a jam!
5. I am building a big dam at the river.
   The dam will keep back the water.
6. Pam makes pills in the lab.
7. She puts on a pin, and it jabs her finger.
8. Did the cab ram into the truck?
9. They will nab the man when he gets there.
10. We get ham from pigs.
11. Pam is keeping tabs on her husband.
12. They are gabbing on the telephone.
13. We keep rats at the lab.
14. Cabs and trucks are in a jam on this street.
15. Pam and Jim are fishing at the dam.

Practice 8-D: Word Families -ack, -ax

1.

back	wax
Jack	**-ax**
-ack	

2. Write the letter or letters and say the word.

b ____ ack	qu ____ ack
j ____ ack	sh ____ ack
l ____ ack	J ____ ack
p ____ ack	
r ____ ack	ax
s ____ ack	t ____ ax
t ____ ack	w ____ ax
bl ____ ack	M ____ ax

3. Read the words.

tack	pack	tax
sack	black	wax
rack	Jack	ax
shack	back	Max
jack	quack	
lack		

4. Write the word you hear.

1. __________	9. __________
2. __________	10. __________
3. __________	11. __________
4. __________	12. __________
5. __________	13. __________
6. __________	14. __________
7. __________	15. __________
8. __________	

5. **New Words**

pay car

6. Read the sentences.

1. Max will jack up his car.
2. We must pay our tax.
3. Go out to the shack and get the ax.
4. Did she pack her black dress?
5. Max is waxing his car.
6. Put the apples into a big sack.
7. Sam hits Jack on the back. Jack's back hurts.
8. Tack up this picture.
9. Jack will pay for the black car.
10. She puts the dress back on the rack.
11. I went to the shop for a box of tacks.
12. The jack is in the back of the car.
13. She lacks ten cents to pay for the gum.
14. Can you pay me back?
15. The ducks are quacking.
16. Jack gives me a pack of gum.
17. The can of olives is in that sack.
18. Jack put the car on the rack to fix it.

Practice 8-E: Word Families -ag, -ad

1.

bag	bad
tag	dad
-ag	**-ad**

2. Write the letter or letters and say the word.

b	____	ag			ad
g	____	ag	b	____	ad
l	____	ag	d	____	ad
n	____	ag	h	____	ad
r	____	ag	l	____	ad
s	____	ag	m	____	ad
t	____	ag	p	____	ad
w	____	ag	s	____	ad
sh	____	ag			

3. Read the words.

tag	wag	mad	had
gag	shag	bad	dad
lag	rag	pad	lad
bag	sag	ad	
nag		sad	

4. Write the word you hear.

1. ____________	10. ____________
2. ____________	11. ____________
3. ____________	12. ____________
4. ____________	13. ____________
5. ____________	14. ____________
6. ____________	15. ____________
7. ____________	16. ____________
8. ____________	17. ____________
9. ____________	

5. Review Word

seventy

6. Read the sentences.

1. Sam looks at the ads.
2. When Dad nags me, I get mad.
3. The shag rug has a pad under it.
4. They gag the man to keep him from yelling.
5. She is sad that she has no dad.
6. Look at the tags on that dress.
7. The bad man ran up to me. He had a gun!
8. Jack put the rags into a bag.
9. Your dress sags in the back.
10. Dan had bad cuts on his hands.
11. The ad says a bag of apples is seventy cents.
12. Kitty tags her sister on the back.
13. Pam rubs the wet dish with a rag.
14. A pen and a pad are by the telephone.
15. Did she pack her bags?
16. Mother gets mad if Dad just sits.
17. Jan is sad that the lad had to go.
18. Do not pat the cat when it is mad.

Practice 9-A: Word Families -ap, -as, -ass

1.

cap	gas	glass
map	**-as**	**-ass**
-ap		

2. Write the letter or letters and say the word.

c ____ ap	g ____ as
g ____ ap	
l ____ ap	b ____ ass
m ____ ap	m ____ ass
n ____ ap	p ____ ass
s ____ ap	gl ____ ass
t ____ ap	gr ____ ass
y ____ ap	
ch ____ ap	

3. Read the words.

nap	gas
tap	
gap	mass
chap	grass
map	bass
lap	glass
yap	pass
cap	
sap	

4. Write the word you hear.

1. ____________	9. ____________
2. ____________	10. ____________
3. ____________	11. ____________
4. ____________	12. ____________
5. ____________	13. ____________
6. ____________	14. ____________
7. ____________	15. ____________
8. ____________	

5. Review Words

happy laugh basket wind

New Words

ask where

6. Read the sentences.

1. The children are taking a nap.
2. Where is that city on the map?
3. I have on my red cap.
4. I ask her, "Can you pass that car?"
5. Wind and water can chap your hands.
6. I am happy my car did not run out of gas.
7. We laugh at the pup when it yaps.
8. The black cat jumps out of my lap.
9. I ask him to pass me a glass of water.
10. Our family will go to mass at ten.
11. There are gaps where things are missing.
12. He gets up from his nap to cut the grass.
13. "Where do you go to fish for bass?" I ask.
14. He puts the gas cap back on the car.
15. I look at the map on my lap.
16. The children tap on the glass.
17. They will not let you in unless you have a pass.
18. He is happy to get this basket of apples.

Practice 9-B: Word Families -ang, -ank

1.

rang	thank
sang	sank
-ang	**-ank**

2. Write the letter or letters and say the word.

b ____ ang	b ____ ank
f ____ ang	r ____ ank
g ____ ang	s ____ ank
h ____ ang	t ____ ank
r ____ ang	y ____ ank
s ____ ang	th ____ ank
	H ____ ank

3. Read the words.

hang	yank
rang	bank
sang	thank
gang	Hank
fang	tank
bang	rank
sank	

4. Write the word you hear.

1. ____________	8. ____________
2. ____________	9. ____________
3. ____________	10. ____________
4. ____________	11. ____________
5. ____________	12. ____________
6. ____________	13. ____________
7. ____________	

5. Review Word

half

Word from skill book 1

number

6. Read the sentences.

1. Thanks for taking the check to the bank.
2. The telephone rang.
3. Hank is in a big gang of kids.
4. The snake has two big fangs.
5. Pam hangs up her dress.
6. The children sang and rang the bells.
7. Jan puts some fish in the tank.
8. The ship sank into the river.
9. Hang up that picture of our gang.
10. Hank yanks the bag out of my hands.
11. Thanks for cutting the grass.
12. Hank ranks number one in hits and runs.
13. A bell rang as the gang ran out of the bank.
14. I have half a tank of gas.
15. Jan cut her bangs.
16. Many pictures are hanging in the bank.
17. She sang to her kids when she put them to bed.
18. She bangs the pans when she is mad.

Practice 9-C: Word Families -ash, -atch, -ath

1.

cash	catch	path
mash	match	math
-ash	**-atch**	**-ath**

2. Write the letter or letters and say the word.

	ash	c	____	atch
c	____ ash	h	____	atch
d	____ ash	l	____	atch
g	____ ash	m	____	atch
h	____ ash	p	____	atch
l	____ ash			
m	____ ash	b	____	ath
r	____ ash	m	____	ath
s	____ ash	p	____	ath

3. Read the words.

dash	hatch
mash	patch
rash	catch
ash	match
hash	latch
sash	
lash	path
cash	bath
gash	math

4. Write the word you hear.

1. ____________	10. ____________
2. ____________	11. ____________
3. ____________	12. ____________
4. ____________	13. ____________
5. ____________	14. ____________
6. ____________	15. ____________
7. ____________	16. ____________
8. ____________	17. ____________
9. ____________	

5. Review Words

after factory

New Word

work

6. Read the sentences.

1. After he works at the factory, he takes a bath.
2. Jan has a red rash on her hands.
3. He lit a match.
4. My pup runs up the path after a cat.

   I dash after him, but I cannot catch him.
5. Mash some yams to have with the hash.
6. Dan likes to work with numbers and do math.

   He works at the bank cashing checks.
7. The chicks will hatch from the eggs.
8. Her red hat does not match her red dress.
9. If I can catch that pup, I will give it a bath.
10. Patch the rip in the tan dress.
11. The glass cut a big gash in his hand.
12. That dress has a pretty sash with it.

    Will you pay cash for the dress?
13. He wants to catch some fish.
14. Sam ran up the path to the glass factory.
15. The latch will not catch.

Practice 9-D: Adding Endings

1. Adding *-y* to words

wind + y → windy

Bill ____________

itch ____________

hand ____________

mess ____________

puff ____________

2. Label the last three letters. Then write the word with *-y*.

VCC + y

jump + y → jumpy

mush ____________

hill ____________

dress ____________

luck ____________

fish ____________

3. Label the last three letters. Then write the word with *-y*.

CVC + C + y

Ken + n + y → Ken-ny

gum ____________

bag ____________

nut ____________

fun ____________

Peg ____________

4. Label the last three letters with VCC or CVC. Then write the word with *-y*.

Tim ____________

sun ____________

grass ____________

dad ____________

bud ____________

fuss ____________

5. Add *-y* and read the word. Double the last letter if you have to.

Pat ____________

glass ____________

Dan ____________

pup ____________

chunk ____________

Sam ____________

Ted ____________

mud ____________

6. Take off *-y* and write the root word.

lucky ____________

chilly ____________

witty ____________

puffy ____________

runny ____________

messy ____________

handy ____________

chatty ____________

7. Reviewing endings

Circle all the words that are the same as the first one.

puff	puffs	puff	puffing	puff	puffy	puffs	puff
bags	bag	baggy	bags	bagging	bags	baggy	bag
chilly	chilly	chills	chilling	chilly	chill	chills	chilling
tap	taps	tap	tapping	taps	tap	tapping	tap
picking	pick	picks	picking	picky	picks	picking	picky

Practice 9-E: Contrasting Short Vowels a and e

a	**bag**	**fad**	**pack**	**sat**	**than**	**lad**
e	**beg**	**fed**	**peck**	**set**	**then**	**led**

1. Write the letter.

1. bat
   bet
   b ____ t

3. ham
   hem
   h ____ m

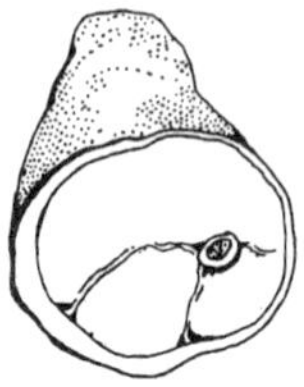

5. tan
   ten
   t ____ n

2. shall
   shell
   sh ____ ll

4. bag
   beg
   b ____ g

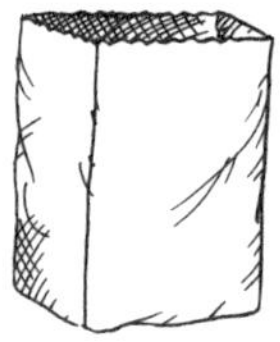

2. Circle the word that you hear.

1. mass mess	4. fad fed	7. bad bed	10. lad led	13. ham hem
2. lag leg	5. bag beg	8. shall shell	11. pack peck	14. vat vet
3. pat pet	6. pan pen	9. sat set	12. than then	15. lass less

3. Write the short vowel that you hear.

1. ch ____ ck	4. p ____ d	7. d ____ b	10. j ____ m	13. t ____ x
2. sh ____ d	5. ____ dge	8. f ____ ll	11. l ____ d	14. h ____ n
3. b ____ t	6. r ____ ck	9. w ____ b	12. b ____ ck	15. ____ d

4. Circle the right word.

1. The (pan / pen) has black ink in it.

2. I (sat / set) out in the sun to get a (tan / ten).

3. Mother (fad / fed) the kids at six.

4. He makes less (than / then) I do.

5. She has a (bad / bed) cut on her leg.

6. We (shall / shell) take the sick cat to a (vat / vet).

Practice 9-F: Contrasting Short Vowels **a** and **i**

a	**dad**	**lack**	**ham**	**fan**	**bat**	**nap**
i	**did**	**lick**	**him**	**fin**	**bit**	**nip**

1. Write the letter.

1. wag
   wig
   w ____ g

2. sack
   sick
   s ____ ck

3. lap
   lip
   l ____ p

4. pan
   pin
   p ____ n

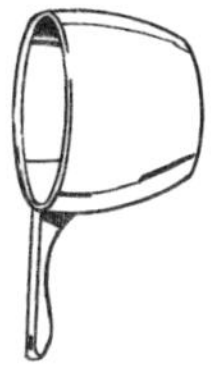

5. bag
   big
   b ____ g

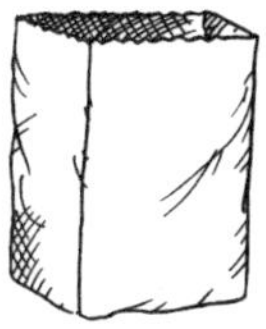

2. Circle the word that you hear.

1.	pack	pick	4.	fat	fit	7.	dash	dish	10.	thank	think	13.	rang	ring
2.	dam	dim	5.	tack	tick	8.	rap	rip	11.	quack	quick	14.	sap	sip
3.	had	hid	6.	bag	big	9.	mass	miss	12.	lad	lid	15.	pan	pin

3. Write the short vowel that you hear.

1. r ____ m	4. t ____ n	7. s ____ ng	10. r ____ nk	13. b ____ d
2. h ____ t	5. l ____ nk	8. l ____ b	11. b ____ n	14. t ____ p
3. s ____ ck	6. h ____ tch	9. p ____ t	12. h ____ s	15. s ____ t

4. Circle the right word.

1. The little girl will take a nap / nip.
2. Put the lad / lid on that.
3. Be quack / quick! Pack your bag / big!
4. The dress has a rap / rip in it.
5. Thank / Think you for the gift.
6. This hat / hit does not fat / fit.

Practice 9-G: Contrasting Short Vowels a and u

a	**ran**	**tag**	**ham**	**rat**	**cab**	**sang**
u	**run**	**tug**	**hum**	**rut**	**cub**	**sung**

1. Write the letter.

1. fan
   fun
   f ____ n

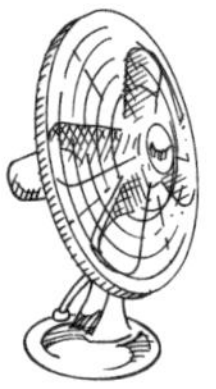

3. hat
   hut
   h ____ t

5. bad
   bud
   b ____ d

2. tab
   tub
   t ____ b

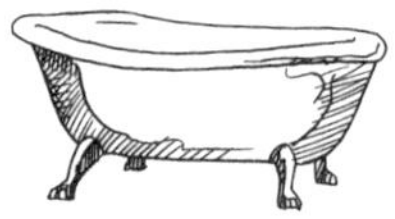

4. tack
   tuck
   t ____ ck

2. Circle the word that you hear.

1.	back	buck	4.	mad	mud	7.	bank	bunk	10.	cap	cup	13.	hang	hung
2.	hag	hug	5.	pan	pun	8.	bat	but	11.	rag	rug	14.	sack	suck
3.	cat	cut	6.	lack	luck	9.	ram	rum	12.	badge	budge	15.	rash	rush

3. Write the short vowel that you hear.

1. n ____ b	4. b ____ n	7. r ____ g	10. l ____ g	13. h ____ sh
2. b ____ zz	5. j ____ g	8. g ____ sh	11. s ____ nk	14. b ____ g
3. m ____ ss	6. c ____ ff	9. p ____ ck	12. s ____ p	15. m ____ tch

4. Circle the right word.

1. Mother yells when she gets mad / mud .
2. Dad gives his little girl a hag / hug .
3. Jan takes a cab / cub to the shop.
4. Put the rags in this sack / suck .
5. Bill must rash / rush to get the bus.
6. Jack works in the bank / bunk .

Practice 9-H: Review Lesson – Short a

1. Look at the picture and say the word. Then fill in the right letter to make the word.

g
sa ____
ck

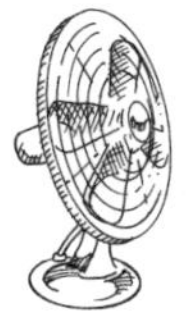

v
____ an
f

t
ca ____
n

b
____ ath
p

l
____ ag
t

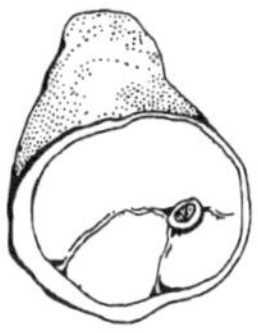

ng
ha ____
m

2. Circle the word that makes sense in the sentence.

1. Bring the fan (back / bag) after you fix it.
2. The fish (sang / sank) in the tank.
3. Dad (rags / nags) at me to do my math.
4. Hank (sat / fat) on the grass by the path.
5. I am (pad / sad) to see Nan go.
6. Jack can pay the (tack / tax). So can his (bad / dad).

3. Look at the picture and say the word. Then circle the right word.

1.  fan van ran vat

2.  tax tank tag tack

3.  fat sat hat that

4. 

cab gab cab nab

5.  bang bank bag ban

6.  grass gas glass pass

Practice 9-H: Review Lesson – Short a (continued)

4. Look at the picture and say the word. Then write the word under the picture.

 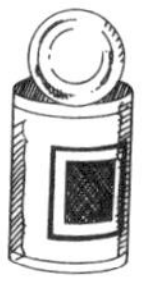 

________ ________ ________

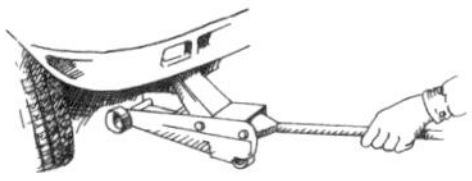  

________ ________ ________

5. Circle all the words that are the same as the first. Work from left to right.

nap	pan	map	nap	rap	nag	nap	nab
sag	sad	sag	gag	sap	gas	wag	sag
mad	map	mad	had	dam	mad	mud	mad
ban	bang	nab	ban	pan	ban	bank	bun
dab	dab	dad	bad	dab	gab	bad	dab
pat	pal	tap	pat	bat	path	pet	pat
mash	mash	math	mash	mass	mash	sham	hash
lag	lap	lag	lad	gal	tag	lag	hag

6. Read these sentences.

1. My van ran out of gas. I will have to take a cab.
2. Jan packs a hat and a tan dress in her bag.
3. Ann Black pays cash when she shops.
   She gets a ham, a can of yams, and a sack of apples.
4. Sam had a chat with the man at the bank.
5. Pam gets mad if the cat jumps into her lap.
6. The gang sang at the mass. Did Pat thank them?
7. I will catch up to Dan's car and then pass it.

Practice 10-A: Word Families -ot, -on

1.

hot	on
not	Don
-ot	**-on**

2. Write the letter or letters and say the word.

c ____ ot	sh ____ ot
d ____ ot	sp ____ ot
g ____ ot	
h ____ ot	on
j ____ ot	D ____ on
l ____ ot	R ____ on
n ____ ot	
p ____ ot	
t ____ ot	

3. Read the words.

dot	got	Ron
pot	shot	Don
cot	lot	on
spot	jot	
hot	tot	
not		

4. Write the word you hear.

1. ____________	8. ____________
2. ____________	9. ____________
3. ____________	10. ____________
4. ____________	11. ____________
5. ____________	12. ____________
6. ____________	13. ____________
7. ____________	14. ____________

5. **Review Word**

building	Chan	doctor	Dr.
head	John	Molly	office

6. Read the sentences.

1. Don does not work a lot.
2. Molly's head is hot.
   She has red spots.
   The doctor will give Molly a shot.
3. Don has two cots in his tent.
4. Pam has on a dress with pink dots.
   Did she get a spot on her dress?
5. It got hot in the doctor's office.
6. Ron does not have any pots and pans.
7. He jots the words on a pad.
8. Ron shot a lot of ducks with his gun.
9. I sing to the little tots.
10. Did Don work out in the hot sun?
11. Dr. Chan says lots of men got sick.
12. That pot is hot. Do not pick it up.
13. She got a letter from John.
14. Ron will put up a building on that lot.

Practice 10-B: Word Family -op

1. top

shop

-op

2. Write the letter or letters and say the word.

c ____ op	st ____ op
h ____ op	sh ____ op
m ____ op	ch ____ op
p ____ op	
t ____ op	
dr ____ op	

3. Read the words.

mop	stop
chop	pop
top	drop
cop	hop
shop	

4. Write the word you hear.

1. ______________________
2. ______________________
3. ______________________
4. ______________________
5. ______________________
6. ______________________
7. ______________________
8. ______________________
9. ______________________

5. Review Words

doll dollar

New Words

every corn

6. Read the sentences.

1. Mop up every drop of water!
2. Stop the car! I want to get out!
3. Do not drop your dishes in the sink.
4. Jim asks Pat to chop up the eggs.
5. At the top of the hill, a girl ran into the street.

   A red car hit the girl, but it did not stop.

   The cops are looking for the car.

   They are stopping every red car in the city.
6. Her hands are wet, and she drops a dish.
7. The cops are keeping the men back.
8. Every kid likes to pop corn.

   The corn pops when it gets hot.

   Keep the top on the pan when the corn pops.
9. She pays six dollars for the doll in the shop.
10. Molly is chopping up nuts for the fudge.
11. After Bill fell, he had to hop on one leg.
12. If I drop this glass of pop, I must mop it up.
13. Ann is shopping for ham, chops, and corn.

Practice 10-C: Word Families -ob, -od

1. Bob rod
job **-od**
-ob

2. Write the letter or letters and say the word.

c _____ ob	c _____ od
j _____ ob	n _____ od
m _____ ob	p _____ od
r _____ ob	r _____ od
s _____ ob	G _____ od
B _____ ob	

3. Read the words.

rob	rod
mob	nod
job	cod
sob	God
Bob	pod
cob	

4. Write the word you hear.

1. _____________	7. _____________
2. _____________	8. _____________
3. _____________	9. _____________
4. _____________	10. _____________
5. _____________	11. _____________
6. _____________	

5. Review Words

Tom lock

New Word

Mom

Word from skill book 1

Robert

6. Read the sentences.

1. Robert wants some corn on the cob.
2. Mom will hang the wet socks on racks.
3. Robert nods his head at Mom.
4. Bob is looking at TV.
5. Mom tells her children that God is good.

   She tells them to thank God.
6. Kim will give him a fishing rod.

   He wants to fish for cod.
7. A man robs the bank.
8. Bob got a job in the city.

   Then Tom said to Bob, "Help me get a job."
9. Pam is sad. She is sobbing.
10. There is a mob at the factory.

    The men and women want jobs.
11. We are having corn on the cob for dinner.
12. Bob asks Pat, "Will you go with me?"

    Pat nods her head.
13. Tom locks the shop so that no one can rob it.

Practice 11-A: Word Families -ock, -ox

1. lock box
   rock **-ox**
   -ock

2. Write the letter or letters and say the word.

d ____ ock	ox
l ____ ock	b ____ ox
m ____ ock	f ____ ox
r ____ ock	
s ____ ock	
cl ____ ock	
sh ____ ock	

3. Read the words.

rock	box
lock	ox
shock	fox
dock	
mock	
clock	
sock	

4. Write the word you hear.

1. __________	6. __________
2. __________	7. __________
3. __________	8. __________
4. __________	9. __________
5. __________	10. __________

5. **Review Words**

better father said skirt
stopped was were

New Word

o'clock

6. Read the sentences.

1. Pat will get a shock if you tell her that!
2. The kids pick up pretty rocks at the river.
3. He ran by as quick as a fox.
4. The ship will be at the dock at ten o'clock.
5. Jill put the skirts and socks in the box.
6. They were sitting on a big rock by the river.
7. Is there a clock in the kitchen?
8. The fox got the hen. I shot at the fox.
9. That man is as big as an ox.
10. Don locks up the shop at five o'clock.
11. Father said to his little son, "Put your socks on."
12. I got a shock when I was fixing the TV.
13. Jan rocks her little girl.
14. My clock has stopped. Is it six o'clock yet?
15. Tom can box better than Bob.
16. This lock works better than the other one.

Practice 11-B: Adding -ed or -d to Action Words to Show Past Action

1. Adding *-ed* to VCC words

VCC + e d

w a n t + e d ⟶**wanted**

b u r n + e d ⟶**burned**

h e l p + e d ⟶**helped**

Add *-ed* and read the word.

work ____________________

jump ____________________

ask ____________________

hunt ____________________

2. Adding *-ed* to CVC words

CVC + C + e d

c h a t + t + e d ⟶ **chatted**

b e g + g + e d ⟶ **begged**

s h o p + p + e d ⟶ **shopped**

Add *-ed* and read the word.

sob ____________________

tap ____________________

hop ____________________

spot ____________________

3. Adding *-d* to words that end in *-e*

l i v e + d ⟶ **lived**

l i k e + d ⟶ **liked**

Add *-d* and read the word.

judge ____________________

whistle ____________________

nudge ____________________

live ____________________

4. Add *-d* or *-ed* and read the words.
Double the last letter if you have to.

pass ____________________

tap ____________________

like ____________________

rock ____________________

chat ____________________

burn ____________________

yank ____________________

pat ____________________

5. Add *-ed* or *-d* to the word and fill the blank. Double the last letter if you have to. Read the sentence.

1. check I ____________ my work.
2. live We ____________ on a hill.
3. lock Molly ____________ the box.
4. nod Father ____________ his head.
5. like Jack ____________ his work.
6. rob Two men ____________ the shop.
7. hatch The chicks ____________.
8. rub Mom ____________ my back.
9. pass That car ____________ me.
10. stop The cop ____________ the car.
11. can Mom ____________ the corn.
12. mess They ____________ up the kitchen.
13. chop I ____________ the olives.
14. shock Bob ____________ me.
15. beg The pup sat up and ____________.
16. match Her dress ____________ her bag.

Practice 11-C: Contrasting Short Vowels a and o

a	**an**	**cab**	**sack**	**tap**	**hat**
o	**on**	**cob**	**sock**	**top**	**hot**

1. Write the letter.

1. map
   mop
   m ____ p

3. pat
   pot
   p ____ t

5. ax
   ox
   ____ x

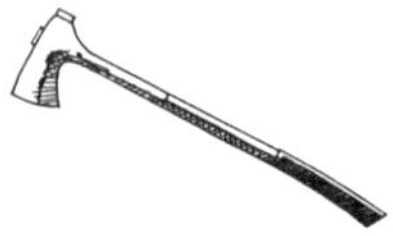

2. rack
   rock
   r ____ ck

4. cab
   cob
   c ____ b

2. Circle the word that you hear.

1. rat	rot	4. shack	shock	7. cat	cot	10. mad	mod	13. lack	lock
2. cap	cop	5. jab	job	8. tap	top	11. ax	ox	14. Dan	Don
3. pad	pod	6. sad	sod	9. rack	rock	12. hat	hot	15. chap	chop

3. Write the short vowel that you hear.

1. c____n	4. m____b	7. h____b	10. p____t	13. s____ck
2. H____nk	5. t____ck	8. r____d	11. r____n	14. s____p
3. l____p	6. c____sh	9. t____b	12. t____t	15. m____d

4. Circle the right word.

1. The cap / cop runs after the men.

2. I have a good jab / job at the factory.

3. A red dress is hanging an / on the rack / rock.

4. Jan is happy, not sad / sod.

5. Your cat / cot is sitting on my hat / hot.

6. He will chap / chop that with an ax / ox.

Practice 11-D: Contrasting Short Vowels u and o

u	**cub**	**hut**	**duck**	**rut**	**pup**
o	**cob**	**hot**	**dock**	**rot**	**pop**

1. Write the letter.

1. luck
   lock
   l ____ ck

3. cup
   cop
   c ____ p

5. cut
   cot
   c ____ t

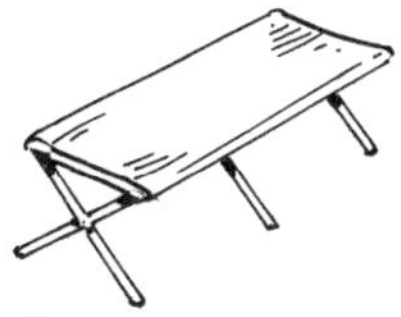

2. nut
   not
   n ____ t

4. suck
   sock
   s ____ ck

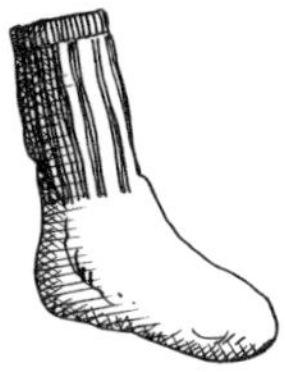

2. Circle the word that you hear.

1. nut	not	4. hut	hot	7. run	Ron	10. pup	pop	13. duck	dock
2. shuck	shock	5. cut	cot	8. mud	mod	11. jut	jot	14. cub	cob
3. rub	rob	6. sub	sob	9. shut	shot	12. hunk	honk	15. suck	sock

3. Write the short vowel that you hear.

1. m____m	4. b____t	7. j____g	10. m____p	13. p____t
2. t____ck	5. j____t	8. r____t	11. h____t	14. c____p
3. g____t	6. r____ck	9. d____ck	12. r____n	15. ch____ck

4. Circle the right word.

1. That man wants to rub / rob the shop.
2. Dr. Hill will give Tom a shut / shot .
3. The pup / pop jumps into the mud.
4. It was hut / hot in the kitchen.
5. Put the cup / cop in the sink.
6. He does nut / not have good luck / lock .

Practice 11-E: Review Lesson – Short o

1. Look at the picture and say the word. Then fill in the right letter to make the word.

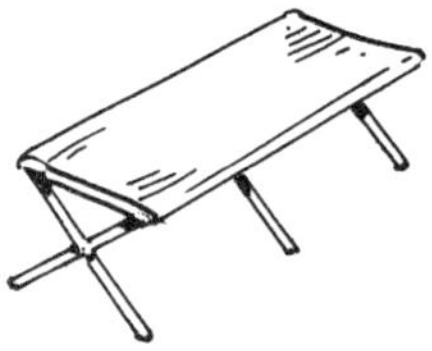

g
____ ot
c

b
mo ____
p

r
____ ock
l

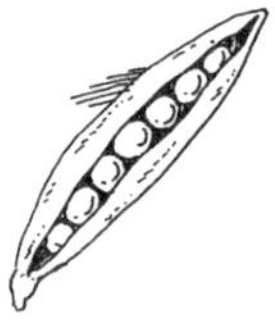

d
po ____
t

b
co ____
d

ch
____ op
sh

2. Circle the word that makes sense in the sentence.

1. We must stop / spot the car!
2. We will have corn on the cob / cop.
3. The lots / tots take naps at two o'clock.
4. What a shock! Bob is quitting his jot / job!
5. Get rid of the robbing / rotting apples.
6. Tom works at his dad's shot / shop.

3. Look at the picture and say the word. Then circle the right word.

1.  chop mop hop shop
2.  lock rock sock lot
3.  dot spot pod pot
4.  fox box pox pot
5.  cob cod cop cot
6.  hot lot not rot

4. Look at the picture and say the word. Then write the word under the picture.

5. Circle all the words that are the same as the first. Work from left to right.

got	got	jot	tot	get	pot	got	cot
pot	cot	pot	jot	pet	pot	got	pot
top	pot	stop	top	tot	top	hop	pot
god	gob	god	dot	pod	god	cod	god
lock	lock	lick	clock	lock	lack	luck	lock
spot	pot	stop	spot	shot	pots	spot	stop
box	fox	ox	box	ox	box	fox	fox
nod	rod	nod	don	nod	not	mod	nod

6. Read these sentences.

1. Ron pays half a dollar for a can of pop.
2. Bob is at the dock. He is fishing with his rod.
3. Don has not had the top job.
4. John shot at the fox.
5. Ron does not lock up the cash box in the shop.
   A man robs the shop.
6. The mob was mad at the cops.
7. Molly drops a cup of water. Mom mops it up.

Practice 12: Adding -es to Words

1. | -s
-x
-sh
-ch | + es, not s |
|---|---|

k i s s + e s → kisses
f i x + e s → fixes
d i s h + e s → dishes
m a t c h + e s → matches

2. ***-s* Words**
Add *-es* to these words.
Fill in the blanks. Read the sentences.

1. kiss Jim ____________ his girl.
2. guess I will give you two _________.
3. fuss Ed _________ when he is sick.
4. mess My son _________ up the tent.
5. glass I like my pink _________.
6. dress Pam gets two new _________.
7. pass A big car _________ my car.

3. ***-x* Words**
Add *-es* to these words.
Fill in the blanks. Read the sentences.

1. ax We have two ________ in the shed.
2. box She sent me two _______ of fudge.
3. fix Jan ____________ a duck dinner.
4. tax We pay so many ____________!
5. fox The __________ ran after the hens.
6. mix Meg ____________ the chip dip.
7. wax Dan ____________ his new truck.

4. ***-sh* Words**
Add *-es* to these words.
Fill in the blanks. Read the sentences.

1. dash Kim ________ after the pup.
2. wish She ________ for a black car.
3. ash The shed burns to ________.
4. rush Mom ________ out of the kitchen.
5. cash Tom ________ his check.
6. dish Jack brings the ________ to the sink.
7. mash Nan ________ the yams.
8. fish Bud ________ in the river.

5. ***-ch* Words**
Add *-es* to these words.
Fill in the blanks. Read the sentences.

1. pitch Fred _________ to us.
2. hatch The chick _________ out of the egg.
3. ditch The men are digging big ________.
4. match Take the _________ from the kids.
5. rich He wants _________.
6. patch Bess puts some ______ on her skirt.
7. catch Ken _________ the pup.
8. itch Chuck's hand _________.

6. Take off the -es and write the root word.
Read the words.

catches ____________________
rushes ____________________
mixes ____________________
passes ____________________
ditches ____________________
wishes ____________________
taxes ____________________
kisses ____________________
rashes ____________________

Practice 13-A: Adding -er or -r to Words

1. Adding *-er* to VCC words

V C C + e r
t h i c k + e r ⟶ thicker

Add *-er* and read the words.

rich ________________

black ________________

2. Adding *-er* to CVC words

C V C + C + e r
t h i n + n + e r ⟶ thinner

Add *-er* and read the words.

hot ________________

big ________________

3. Just add *-r* to words that end with *-e.*

l a r g e + r ⟶ larger

Add *-r* and read the words.

little ________________

large ________________

4. Add *-er* or *-r* to the word. Double the last letter if you have to. Read the word.

fresh ________________

wet ________________

large ________________

sick ________________

red ________________

dark ________________

sad ________________

quick ________________

rich ________________

little ________________

5. Add *-er* or *-r* to the word and fill in the blank. Double the last letter if you have to. Read the sentence.

1. rich — He is ________ than I am.
2. hot — It is getting much ________.
3. mad — He got ________ when I stopped him.
4. large — This park is ________ than the other one.
5. thin — This kid is ________ than that kid.
6. dark — Your dress is a ________ red than my dress.
7. fresh — This egg is ________ than that egg.
8. wet — Take this rag. It is ________ than that one.
9. big — John is ________ than his brother.
10. hard — This work is much ________ to do.
11. quick — The cat is ________ than the rat.
12. up — I had a nap on the ________ bunk.
13. thick — This mix is getting ________.
14. fat — Dad is ________ than Mom.
15. red — Jill is getting ________ in the sun.
16. sick — Tom was ________ than his brother.

Practice 13-B: Adding -er or -r to Action Words

1. Adding *-er* or *-r* to action words

This part <u>starts</u> the car.
It is a <u>starter</u>.

Jack <u>sits</u> with kids.
He is a <u>sitter</u>.

Pat <u>writes</u> for the paper.
She is a <u>writer</u>.

2. Adding *-er* or *-r* to action words

VCC+ er
s t a r t + e r ⟶ starter

CVC+C+er
s i t + t + e r ⟶ sitter

+ r
w r i t e + r ⟶ writer

3. Add *-er* or *-r* to the words. Double the last letter if you have to. Read the words.

sell ____________
hit ____________
help ____________
make ____________
run ____________
build ____________
farm ____________
win ____________
think ____________
whistle ____________

4. Add *-er* or *-r* to the word. Double the last letter if you have to. Read the sentence.

1. sing — Billy wants to be a ____________.
2. work — The factory ____________ likes her job.
3. win — Peg is the ____________!
4. bat — The ____________ hurt his arm.
5. write — That ____________ is very witty.
6. rock — Jan is sitting on the ____________.
7. rob — The cop will catch the ____________.
8. bid — He sells the TV to the top ____________.
9. bank — The ____________ looks at the checks.
10. kill — Get some bug ____________ at the shop.

5. Some other words have double letters and *-er*. But the *-er* is part of the word. It is not an ending that is added.
Read the words. Use them to fill in the blanks. Read the sentences.

1. suffer — If pets do not have water, they ____________.
2. copper — They make the pans out of ____________.
3. ladder — The man is standing on a ____________.
4. butter — I put half a cup of ____________ in the fudge.
5. hammer — Hit the tack with a ____________.
6. dinner — We are having ham for ____________.
7. letter — Jim writes Pat a ____________.
8. litter — Pick that up! Do not ____________.
9. matter — It does not ____________ when he comes.
10. summer — It gets very hot in the ____________.

Practice 13-C: Endings for Words that End with Consonant + y

1. Adding *-ing*

Just add *-ing.*

c a r r y + i n g ⟶ carrying

Add *-ing* to the word and fill in the blank. Read the sentence.

Take off the ending. Write the root word.

carry	The men are ________ a TV set.		hurrying	________
hurry	Dr. Chan is ________ to her office.		carrying	________

2. Adding *-ed*

Change *y* to *i,* and then add *-ed.*

m a r r y

m a r r i + e d ⟶ married

Add *-ed* to the word and fill in the blank. Read the sentence.

Take off the ending. Write the root word.

marry	My sister ________ Ken.		carried	________
carry	He ________ the pup to the vet.		married	________

3. Adding *-es* to action words

Change *y* to *i,* and then add *-es.*

h u r r y

h u r r i + e s ⟶ hurries

Add *-es* to the word and fill in the blank. Read the sentence.

Take off the ending. Write the root word.

hurry	Peg ________ to work.		carries	________
carry	Rick ________ the sack on his arm.		hurries	________

4. Adding *-es* to show there is more than one

Change *y* to *i,* and then add *-es.*

c i t y

c i t i + e s ⟶ cities

Add *-es* to the word and fill in the blank. Read the sentence.

Take off the ending. Write the root word.

puppy	Jill has six little ________!		factories	________
lily	The garden shop sells ________.		families	________
jelly	Pam sells jams and ________.		cities	________

5. Adding *-er*

Change *y* to *i,* and then add *-er.*

l u c k y

l u c k i + e r ⟶ luckier

Add *-er* to the word and fill in the blank. Read the sentence.

Take off the ending. Write the root word.

lucky	John was ________ than I was.		messier	________
happy	I am ________ if you are with me.		prettier	________
pretty	Liz is ________ than her sister.		sunnier	________

Practice 14-A: Compound Words

1. Look at these compound words:

sunset	sun	set	forget	for	get	checkout	check	out
anyone	any	one	dishrag	dish	rag	backpack	back	pack

2. Put the two words together to make a compound word. Read the word.

red	head	________________	hand	out	________________
stand	by	________________	pot	luck	________________
street	car	________________	back	lash	________________
up	keep	________________	any	where	________________
bath	tub	________________	patch	work	________________
cat	nip	________________	sun	bath	________________

3. Find the two smaller words in each compound word. Write them in the blanks. Read the smaller words and the compound words.

jackpot	________	________	checkup	________	________
everyone	________	________	makeup	________	________
suntan	________	________	comeback	________	________
outfit	________	________	dugout	________	________
dishwater	________	________	pigpen	________	________
catfish	________	________	something	________	________

4. Make compound words. Match each word on the left with a word on the right. Write the compound word in the blank. The first one is done for you.

pop	popcorn	work	quarter	________________	fish
bath	________________	corn	shell	________________	back
net	________________	what	pad	________________	out
some	________________	tub	sun	________________	lock
for	________________	hill	drop	________________	stand
up	________________	give	under	________________	burn

Practice 14-B: Two-Syllable Words

1. The big words below have been divided into syllables. Using what you have learned about sounding out words, sound out the syllables and figure out each word.

public	/**pub** lic/	rabbit	/**rab** bit/
model	/**mod** el/	admit	/ad **mit**/

2. Read the word first. Then read the sentence.

1. rabbit /**rab** bit/
   The rabbit is in the garden.
2. sudden /**sud** den/
   Did the bus make a sudden stop?
3. public /**pub** lic/
   Anyone can come to a public park.
4. napkin /**nap** kin/
   He has a napkin on his lap.
5. model /**mod** el/
   Peg will model the pretty dress.
6. limit /**lim** it/
   Jim's car is going less than the limit.
7. discuss /**dis** cuss/
   They met to discuss it.

3. Read the word. Fill in the blank. Read the sentence.

1. happen /**hap** pen/
   Did that _______________ at work?
2. picnic /**pic** nic/
   I will carry the _______________ basket.
3. witness /**wit** ness/
   The _______________ looks at the robber.
4. fabric /**fab** ric/
   I cannot get the spot out of the _______________.
5. finish /**fin** ish/
   You can go after you _______________ the work.
6. habit /**hab** it/
   She sucks on her finger. It is a bad _______________.
7. admit /ad **mit**/
   The robber did not _______________ that he did it.

4. A Working Woman

Review word: heard
New word: love

Nan was up on a ladder with a hammer in her hand.
She heard a car stop.
It was her best friend Jim, coming to pick her up from work.

"It is so hot I want to go to the park for a picnic dinner," Nan said.
"This job is hard work, but I have to admit the pay is good."

Jim did not like to discuss jobs and pay.
He did not have a job yet, but he was looking for one.
He liked to fix cars and trucks.

"Shut up!" he said. "It makes me mad to think you will have to pay the bills if we get married!"

Nan stopped her work and hugged Jim.
Nan said, "You will get a job! You are better at fixing cars than anyone!"

Jim hugged her and kissed her.
"Forgive me for getting so mad," he said.
"It just upsets me to think you are working and I have no job."

Nan said, "That is the limit! If I have no job and you are working, do you think I will not love you then? Loving you is a good habit.
I do not want to stop!"

Jim laughed. "I do love you so!" he said.
"You make everything so funny! I am lucky that you love me.
Does it matter if I make less than you?
I will make it up to you. When I have a good job, you can just be with our kids."

"If I want to?" asked Nan.
"If you want to," said Jim.

Practice 1-A: Review of Beginning Digraphs sh- and ch-

1.

sh-	ch-
shop	children
she	chop

2. What does the word begin with?

sh ch

1.	______	9.	______
2.	______	10.	______
3.	______	11.	______
4.	______	12.	______
5.	______	13.	______
6.	______	14.	______
7.	______	15.	______
8.	______	16.	______

3. Write the letters and say the word.

sh	ch
______ ack	______ at
______ ed	______ eck
______ ell	______ ess
______ ip	______ ick
______ ock	______ ill
______ op	______ in
______ ot	______ ip
______ ut	______ op

4. Read the words.

shot	chill
shock	chip
shop	chick
shut	chin
shack	chat
ship	chop
shell	chess
shed	check

5. Write the word you hear.

1.	__________	9.	__________
2.	__________	10.	__________
3.	__________	11.	__________
4.	__________	12.	__________
5.	__________	13.	__________
6.	__________	14.	__________
7.	__________	15.	__________
8.	__________	16.	__________

6. Read the sentences.
Review words: what, shall

1. Do not get a chill out there.
2. The children shop for a chess set. Dan will write a check to pay for it.
3. Shell the nuts and chop them up.
4. That family lives in a little shack. There is a shed in back of the shack.
5. Ed checks the deck of the ship.
6. I shipped the gift when I got the check.
7. Ann chats with a woman in the shop.
8. The hen is with her chicks.
9. What a shock! A man robbed the shop!
10. Shall I shut up the shop?
11. Ted fell and hit his chin.
12. The cup has a chip in it.
13. The hurt man is going into shock. Dr. King will give him a shot.

Practice 1-B: Review of Beginning Digraphs th- and wh-

1. **th-**

thank	th
	____ ank
	____ ick
	____ in
	____ ing
	____ ink

th-

this	th
the	____ an
they	____ at
there	____ em
	____ en
	____ is
	____ us

wh-

whistle	wh
when	____ en
where	____ ip
what	____ ich

2. Read the words.

thin	then	which
thick	them	whip
thing	thus	when
think	that	
thank	than	
	this	

3. Write the word you hear.

1. ____________	8. ____________
2. ____________	9. ____________
3. ____________	10. ____________
4. ____________	11. ____________
5. ____________	12. ____________
6. ____________	13. ____________
7. ____________	14. ____________

4. Read the sentences.
New word: their
Review words: where, what, happen

1. Whip this mix. Then it will get thick.
2. Where do they live? They live there.
3. Which rug shall we get?
   A thick one is better than a thin one.
4. Thanks for thinking of me.
5. When Bob is ten, then Ted will be six.
6. What is that thing in the grass?
   Is it a whip? No, I think it is a snake.
7. What happened to their son?
8. That family has no cash in the bank.
   Thus, they cannot get their car fixed.
9. I will thank them for their gifts.
10. Jan is thinner than her sister.
11. Where do you think the kids went?
12. Which car is best? Which one do you want?
13. What things will you tell them?

Practice 2-A: Review of Ending Digraph -ch

1. **-ch**

catch

much

2. What does the word end with?

ch **sh**

1. _______	9. _______
2. _______	10. _______
3. _______	11. _______
4. _______	12. _______
5. _______	13. _______
6. _______	14. _______
7. _______	15. _______
8. _______	16. _______

3. Write the letters and say the word.

ch	ch
bat _______	hit _______
cat _______	pit _______
hat _______	wit _______
lat _______	not _______
mat _______	ri _______
pat _______	whi _______
it _______	mu _______
dit _______	su _______

4. Read the words.

pitch	catch
ditch	hatch
itch	match
witch	batch
hitch	such
notch	much
latch	which
patch	rich

5. Write the word you hear.

1. _______________	9. _______________
2. _______________	10. _______________
3. _______________	11. _______________
4. _______________	12. _______________
5. _______________	13. _______________
6. _______________	14. _______________
7. _______________	15. _______________
8. _______________	16. _______________

6. Read the sentences.
Review words: curtains, dark, four, kitchen

1. I will pitch and you can catch.
2. Jack's hands are red. They itch.
3. Bill is such a good singer.
4. My kitchen is much bigger than yours.
5. If Ed wins the match, he will be rich.
6. Mom put a patch on the rip in my dress.
7. We want to catch a big batch of fish.
8. Dan is rich. He has four cars.
9. Which eggs will hatch?
10. The curtains do not match the rug.
11. Pam digs a ditch for the water.
12. The latch lifts into this notch.
13. The witch has a black cat with her.
14. John lit a match in the dark.
15. Is there a hitch on your car?
16. Which factory makes such good cars?

Practice 2-B: Review of Ending Digraphs -sh and -th

1. **-sh** **-th**

fish path

2. What does the word end with?

s sh	t th
1. _______	1. _______
2. _______	2. _______
3. _______	3. _______
4. _______	4. _______
5. _______	5. _______
6. _______	6. _______
7. _______	7. _______
8. _______	8. _______

3. Write the letters and say the word.

sh	sh	th
a _______	ma _______	ba _______
ca _______	ra _______	ma _______
da _______	di _______	pa _______
fi _______	wi _______	
ga _______	hu _______	
ha _______	ru _______	
la _______		

4. Read the words.

wish	rash
dish	ash
fish	hash
rush	gash
hush	lash
dash	path
mash	math
cash	bath

5. Write the word you hear.

1. _______________	9. _______________
2. _______________	10. _______________
3. _______________	11. _______________
4. _______________	12. _______________
5. _______________	13. _______________
6. _______________	14. _______________
7. _______________	15. _______________
8. _______________	16. _______________

6. Read the sentences.
Review words: fresh, Smith, burn

1. Fill the dish with nuts.
2. They rush to the bank to cash a check.
3. Fred has a red rash on his hands.
4. Ed takes a bath after his factory work.
5. Can you help Ted with his math?
6. The pups dash up the path.
7. My kids rush into the kitchen, yelling. I tell them to hush.
8. Ted Smith is fixing hash for dinner.
9. I wish we had some fresh fish.
10. The glass cut a gash in her arm.
11. We wish to pay cash for the TV set.
12. Lash the bags to the top of the car.
13. It burned to black ashes.
14. Mash the yams for dinner.
15. Fish is the dish that Mom makes best.

Practice 3: bl-

1. **bl-**

lock	less
block	bless

2. What does the word begin with?

b **l** **bl**

1. ______ 9. ______
2. ______ 10. ______
3. ______ 11. ______
4. ______ 12. ______
5. ______ 13. ______
6. ______ 14. ______
7. ______ 15. ______
8. ______ 16. ______

3. Write the letters and say the word.

bl	**bl**
______ ack	______ ock
______ ank	______ ot
______ ed	______ otch
______ ess	______ uff
______ ink	______ ush

4. Read the words.

bless	blink
bled	blush
blot	bluff
blotch	blank
block	black

5. Write the word you hear.

1. ______________ 6. ______________
2. ______________ 7. ______________
3. ______________ 8. ______________
4. ______________ 9. ______________
5. ______________ 10. ______________

6. Read the sentences.
Review word: name

1. Ted fills in the blanks.
2. He says he can win, but he is bluffing.
3. A black family lives in my building.
4. I get red when I blush.
5. Do not blink when I put the drops in.
6. Mom asks God to bless this dinner.
7. Tim's hand got cut. It bled a lot.
8. The gun shot blanks.
9. The black van is blocking our car.
10. Tom blushes when Pam winks at him.
11. Sam blots up the black ink.
12. Write your name in this blank.
13. "God bless you," Dad said to his son.
14. Fred lives two blocks from me.
15. Bob has a red blotch on his neck.
16. My friends are a blessing to me.

Practice 4: cl-

1. cl-

lock — lap

clock — clap

2. What does the word begin with?

c **l** **cl**

1. _______ 9. _______
2. _______ 10. _______
3. _______ 11. _______
4. _______ 12. _______
5. _______ 13. _______
6. _______ 14. _______
7. _______ 15. _______
8. _______ 16. _______

3. Write the letters and say the word.

cl	**cl**
_______ am	_______ ip
_______ ap	_______ ing
_______ ass	_______ ock
_______ ash	_______ ot
_______ ang	_______ ub
_______ ick	_______ uck
_______ iff	_______ utch

4. Read the words.

cliff	class
cling	clap
clip	clang
click	clam
cluck	clash
club	clot
clutch	clock

5. Write the word you hear.

1. _______________ 8. _______________
2. _______________ 9. _______________
3. _______________ 10. _______________
4. _______________ 11. _______________
5. _______________ 12. _______________
6. _______________ 13. _______________
7. _______________ 14. _______________

6. Read the sentences.
New words: paper, shirt
Review word: color

1. Pam clips a picture from the paper.
2. Class starts at ten o'clock.
3. They are clapping for the winner.
4. Bob is in the Garden Club.
5. Do not stand on the edge of the cliff.
6. Ed went to the river to dig for clams.
7. That red shirt clashes with this skirt.
8. The clock stopped at six o'clock.
9. Jim writes for his class paper.
10. The lock shut with a click.
11. Put the paper clips in that dish.
12. My wet shirt clings to my back.
13. The clutch on my car does not work.
14. The hens cluck at the chicks.
15. Some birds make nests on cliffs.
16. This color will clash with my shirt.

Practice 5: fl-

1. fl-

lock lip
flock flip

2. What does the word begin with?

f l fl

1. ______
2. ______
3. ______
4. ______
5. ______
6. ______
7. ______
8. ______
9. ______
10. ______
11. ______
12. ______
13. ______
14. ______
15. ______
16. ______

3. Write the letters and say the word.

fl	**fl**
______ ag	______ ip
______ ap	______ ing
______ at	______ ock
______ ash	______ op
______ ed	______ uff
______ esh	______ ush
______ ick	______ unk

4. Read the words.

flip	flush
fling	fluff
flick	flop
flat	flock
flash	flesh
flag	fled
flap	flunk

5. Write the word you hear.

1. __________
2. __________
3. __________
4. __________
5. __________
6. __________
7. __________
8. __________
9. __________
10. __________
11. __________
12. __________
13. __________
14. __________

6. Read the sentences.
Review word: barn

1. The flag is flapping in the wind.
2. What is that flashing in the dark?
3. Tim is flat on his back in bed.
4. Flip a quarter. If it is heads, I win.
5. A flock of birds went by.
6. The farmer fled from the burning barn.
7. Bill is flunking his math class.
8. Put in the letter and lick the flap.
9. It will not flush until I fix it.
10. Jan can do a back flip.
11. Ned flops on his bed.
12. Whip the eggs until they fluff up.
13. Tom is fat. He has a lot of flesh.
14. Jack flicks on the TV.
15. The street is hilly, not flat.
16. Ann flings her ring back at Jim.

Practice 6: pl-, gl-

1. **pl-** **gl-**

pl-	gl-
lot	lad
plot	glad

2. What does the word begin with?

p l pl	**g l gl**
1. _______	1. _______
2. _______	2. _______
3. _______	3. _______
4. _______	4. _______
5. _______	5. _______
6. _______	6. _______
7. _______	7. _______
8. _______	8. _______

3. Write the letters and say the word.

pl	**gl**
_______ an	_______ ad
_______ ank	_______ ass
_______ edge	Gl enn
_______ ot	_______ um
_______ us	
_______ uck	
_______ ug	
_______ um	

4. Read the words.

plot	glum
pledge	glass
plus	glad
plum	Glenn
plug	
pluck	
plan	
plank	

5. Write the word you hear.

1. _______________	7. _______________
2. _______________	8. _______________
3. _______________	9. _______________
4. _______________	10. _______________
5. _______________	11. _______________
6. _______________	12. _______________

6. Read the sentences.
New word: plenty

1. Plug in the TV set.
2. I cut my hand on some glass.
3. The men had a plot to rob the bank.
   We are glad that the plan did not work.
4. Peg looks glum. She misses Glenn.
5. I have plenty of planks to make a shed.
6. Bill looks good with his glasses on.
7. Glenn says the pledge to the flag.
8. Jim plucks a hen for dinner.
9. Ed plans to pledge ten dollars.
10. Kim puts the plug in the sink.
11. The Hills pick plenty of plums.
12. One plus six is seven.
13. Dad is planning his garden.
    He plans to have corn on this plot.
14. I am glad I have plenty of glasses.

Practice 7: sl-

1. **sl-**

led	lick
sled	slick

2. What does the word begin with?

s **l** **sl**

1. _______	9. _______
2. _______	10. _______
3. _______	11. _______
4. _______	12. _______
5. _______	13. _______
6. _______	14. _______
7. _______	15. _______
8. _______	16. _______

3. Write the letters and say the word.

sl	**sl**
_______ acks	_______ im
_______ am	_______ ip
_______ ap	_______ ing
_______ ash	_______ ob
_______ ang	_______ ot
_______ ed	_______ ug
_______ ick	_______ um
_______ id	_______ ush

4. Read the words.

sled	slum
slot	slush
slob	slug
slam	slip
slash	slid
slap	sling
slang	slick
slacks	slim

5. Write the word you hear.

1. _______________	9. _______________
2. _______________	10. _______________
3. _______________	11. _______________
4. _______________	12. _______________
5. _______________	13. _______________
6. _______________	14. _______________
7. _______________	15. _______________
8. _______________	16. _______________

6. Read the sentences.
New words: never, down

1. There is slush on the street.
   Do not slip on it.
2. Ann's slacks make her look slim.
3. The shop slashed rugs down to a dollar.
4. My shirt is a mess. I look like a slob.
5. Mom never slaps the children.
6. That sled will never carry four of us.
7. Pam puts on her pink slip.
8. Dr. John put Jim's arm in a sling.
9. If I slim down, my slacks will fit.
10. Put the letters in the slot.
11. Jan never lived in the city slums.
12. Don slammed down the telephone.
13. The children slip on the slick street.
14. My doctor never says slang words.
15. Ben slid down the hill on his sled.

Practice 8: Review of Beginning l Blends: bl-, cl-, fl-, gl-, pl-, sl-

1. Look at the picture and say the word. Then write the letters of the beginning blend you hear.

1.

_____**ed**

4.

_____**ock**

2.

_____**am**

5.
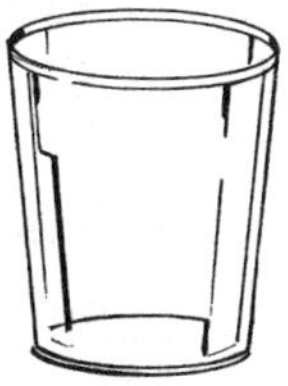

_____**ass**

3.
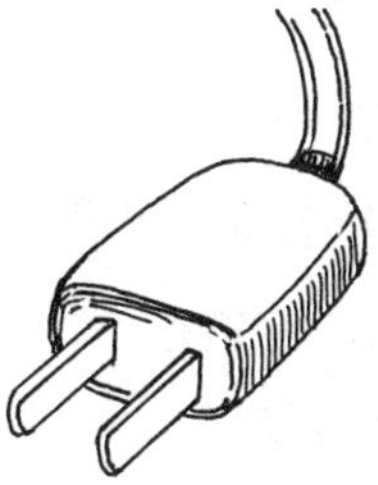

_____**ug**

6.

_____**ag**

2. Practice the blends you have learned.

Make at least 15 words with these beginnings and endings.

bl	ash	_________	_________
cl	ip	_________	_________
fl	ot	_________	_________
pl	ed	_________	_________
sl	um	_________	_________
gl	ing	_________	_________
		_________	_________
		_________	_________

3. Fill in the blends to make sense in the sentences.

pl fl gl bl cl sl

1. Our class says the ____edge to the flag.
2. Glenn is watching a ____ock of birds.
3. There was ____ush on the street. It was ____ick.

   Ed slipped and fell ____at on his back.
4. We are ____ad you plan to visit our club.
5. Tim looks ____um. He is ____unking math.
6. Mom made plenty of ____um jam.
7. His shirt clashes with his ____acks.
8. I have lots of paper clips and ____ank paper.
9. We ____ap for Jan. We make her ____ush.

Practice 9: br-

1. **br-**

ring rag

bring brag

2. What does the word begin with?

b r br

1. ______ 9. ______
2. ______ 10. ______
3. ______ 11. ______
4. ______ 12. ______
5. ______ 13. ______
6. ______ 14. ______
7. ______ 15. ______
8. ______ 16. ______

3. Write the letters and say the word.

br **br**

______ ag ______ ick

Br ad ______ idge

______ an ______ im

______ ass ______ ing

______ at ______ ush

4. Read the words.

bran brush

brat bring

brass brick

Brad bridge

brag brim

5. Write the word you hear.

1. ______ 6. ______
2. ______ 7. ______
3. ______ 8. ______
4. ______ 9. ______
5. ______ 10. ______

6. Read the sentences.
New words: hair, made
Review words: bread, brother, turn

1. Mom brags, "This is good bread I made."

   She made bread with bran in it.
2. Ben brushes his hair quickly.
3. I cannot carry the bricks up that hill.
4. The bridge is slick when it is wet.
5. Fill my glass to the brim.
6. I brushed the cat's hair from my skirt.
7. Kim brings me some jam for the bread.
8. This dish is made of brass.
9. My little brother is a brat.
10. Brad lives in a big brick building.
11. Bring me a glass of water.
12. The brim of his hat was turned down.
13. "My son is a doctor," he brags.
14. The red truck is on the bridge.
15. Brad has to brush up on his math.

Practice 10: cr-

1. **cr-**

rib	rack
crib	crack

2. What does the word begin with?

c **r** **cr**

1. _______
2. _______
3. _______
4. _______
5. _______
6. _______
7. _______
8. _______
9. _______
10. _______
11. _______
12. _______
13. _______
14. _______
15. _______
16. _______

3. Write the letters and say the word.

cr	**cr**
_______ ab	_______ ib
_______ ack	_______ op
_______ am	_______ ush
_______ ash	_______ utch
_______ ank	

4. Read the words.

crush	crank
crutch	crack
crash	crop
cram	crib
crab	

5. Write the word you hear.

1. ______________________________
2. ______________________________
3. ______________________________
4. ______________________________
5. ______________________________
6. ______________________________
7. ______________________________
8. ______________________________
9. ______________________________

6. Read the sentences.
New word: baby

1. The baby is in his crib.
2. Ed got hurt in a car crash.
   A car crushed his leg.
   Ed is on crutches till he gets well.
3. I had to crank up the car to start it.
4. Will you crack the nuts for me?
5. The baby crushed bread in her fingers.
6. Farmers want to have big crops.
7. The brick building fell with a crash.
8. A farmer keeps his corn in a crib.
9. My best pot has a crack in it.
10. When I pack my bags, I cram things in.
    Then my things get crushed.
11. The baby drops the cup and cracks it.
12. We will sell our corn crop.
13. We had crab legs for dinner.

Practice 11: dr-

1. dr-

rip	rug
drip	drug

2. What does the word begin with?

d r dr

1. _______	9. _______
2. _______	10. _______
3. _______	11. _______
4. _______	12. _______
5. _______	13. _______
6. _______	14. _______
7. _______	15. _______
8. _______	16. _______

3. Write the letters and say the word.

dr	**dr**
_______ ag	_______ ink
_______ ank	_______ op
_______ ess	_______ ug
_______ ill	_______ um
_______ ip	_______ unk

4. Read the words.

dress	drank
drop	drag
drum	drip
drunk	drink
drug	drill

5. Write the word you hear.

1. _______________
2. _______________
3. _______________
4. _______________
5. _______________
6. _______________
7. _______________
8. _______________
9. _______________
10. _______________

6. Read the sentences.
New words: school, play

1. Drag the boxes out of the shed.
2. He dropped out of school to play drums.
3. Are the men drilling for water?
4. Sam drank so much he got drunk.
5. Molly's red dress is pretty.
6. Drink plenty of water.
7. That dress is dripping wet.
8. I dropped my fishing rod in the water.
9. She drank every drop of her drink.
10. This school play drags on and on.
11. Water is dripping from the sink.
12. Dr. King locks up the drugs.
13. Don plays the drums after school.
14. She keeps drilling us on our math.
15. We never dress up to go to school.
16. Ed looks drunk, but he is sick.

Practice 12: fr-, pr-

1. **fr-** **pr-**

fr-	pr-
rank	rank
frank	prank

2. What does the word begin with?

f r fr	**p r pr**
1. _______	1. _______
2. _______	2. _______
3. _______	3. _______
4. _______	4. _______
5. _______	5. _______
6. _______	6. _______
7. _______	7. _______
8. _______	8. _______

3. Write the letters and say the word.

fr	**pr**
_______ ank	_______ ank
Fr ank	_______ ess
Fr an	_______ om
Fr ed	_______ op
_______ et	
_______ esh	
_______ ill	

4. Read the words.

fret	prop
fresh	prom
Fred	press
Fran	prank
Frank	
frank	
frill	

5. Write the word you hear.

1. _______________	7. _______________
2. _______________	8. _______________
3. _______________	9. _______________
4. _______________	10. _______________
5. _______________	11. _______________
6. _______________	

6. Read the sentences.
Review words: friend, pretty, market

1. Press down on it to make it flat.
2. My friends played a prank on me.
3. Fred will get some fresh eggs for us.
4. Let me be frank. This is what I think.
5. The baby frets for her mother.
6. Prop the sick girl up in bed.
7. Frank is pressing his black slacks.
8. Fran's dress for the prom has frills.
9. Max has a fresh fish market.
10. The cops pressed the mob back.
11. Do not fret if you cannot do it.
12. The rack fell down. Prop it back up.
13. The women are good friends.
14. Fran has such pretty dresses!
15. The paper goes to press at six.
16. Get some fresh bread for dinner.

Practice 13: gr-

1. **gr-**

rip	ram
grip	gram

2. What does the word begin with?

g **r** **gr**

1. ______	9. ______
2. ______	10. ______
3. ______	11. ______
4. ______	12. ______
5. ______	13. ______
6. ______	14. ______
7. ______	15. ______
8. ______	16. ______

3. Write the letters and say the word.

gr	**gr**
______ ab	______ in
______ am	______ ip
______ ass	______ it
Gr eg	______ ub
______ ill	______ udge

4. Read the words.

grin	grudge
grill	grub
grit	grass
grip	grab
Greg	gram

5. Write the word you hear.

1. ______	6. ______
2. ______	7. ______
3. ______	8. ______
4. ______	9. ______
5. ______	10. ______

6. Read the sentences.
New word: grubby
Review word: ready

1. Greg grins when he is happy.
2. I grab my bag and run for the bus.
3. Does Frank like grits?
4. Ed is grubby. He fell in the mud.
5. Pat grabs his hand and grips it hard.
6. Fran will cut the grass.
7. Do not carry a grudge.
8. Digging ditches is grubby work.
9. Mom mops up the mud and grit.
10. "Grub is ready! Come and get it!"
11. The baby grins when Dad laughs.
12. Set up the grill on the grass.
    Greg wants to grill the fish.
13. I got grubby working on my car.
14. I want to grab the jug from him.
    But I cannot get a good grip on it.

Practice 14: tr-

1. **tr-**

rack	rip
track	trip

2. What does the word begin with?

t **r** **tr**

1. _______	9. _______
2. _______	10. _______
3. _______	11. _______
4. _______	12. _______
5. _______	13. _______
6. _______	14. _______
7. _______	15. _______
8. _______	16. _______

3. Write the letters and say the word.

tr	**tr**
_______ ack	_______ ip
_______ ap	_______ ot
_______ ash	_______ uck
_______ ick	_______ udge
_______ im	_______ unk

4. Read the words.

trot	trip
trunk	trim
truck	trick
trudge	
trap	
trash	
track	

5. Write the word you hear.

1. _______________	6. _______________
2. _______________	7. _______________
3. _______________	8. _______________
4. _______________	9. _______________
5. _______________	10. _______________

6. Read the sentences.

1. We want to catch the rats in the trap.
2. Frank runs at the track.
3. Jack plays a trick on Don.
4. Fran will trim Pat's hair.
5. I got a dress for the trip.
   I will pack it in this trunk.
6. Mom put my shirt in the trash.
7. Do not track mud on the rug.
8. The kids trudge up the hill.
   The pup trots after the kids.
9. Trim the fat from the chops.
10. The jack is in the trunk of the car.
11. Do not trip on that trash can.
12. There are car tracks on the grass.
13. She tells her pet to do a trick.
14. That truck is carrying six pigs.

Practice 15: Contrasting r and l

1.

r	l
	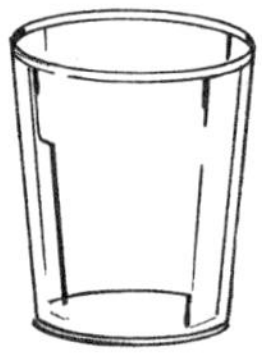
gr	**gl**
grass	glass

2. Look at the picture and say the word. Then circle the blend you hear.

br bl	**cr cl**	**fr fl**	**br bl**	**fr fl**

3. Circle the blend you hear.

1.	fr	fl	11.	br	bl
2.	gr	gl	12.	cr	cl
3.	br	bl	13.	gr	gl
4.	pr	pl	14.	br	bl
5.	cr	cl	15.	fr	fl
6.	gr	gl	16.	cr	cl
7.	cr	cl	17.	br	bl
8.	br	bl	18.	pr	pl
9.	gr	gl	19.	fr	fl
10.	fr	fl	20.	gr	gl

4. Write in *r* or *l* for each word you hear.

1.	c ____ ack	11.	p ____ op
2.	f ____ ock	12.	c ____ ock
3.	p ____ ank	13.	b ____ ush
4.	b ____ ed	14.	g ____ and
5.	g ____ ad	15.	c ____ ass
6.	c ____ amp	16.	c ____ ank
7.	f ____ ank	17.	f ____ esh
8.	p ____ od	18.	g ____ ass
9.	b ____ and	19.	c ____ ap
10.	c ____ utch	20.	c ____ ash

Practice 16: Review of Beginning r Blends: br-, cr-, dr-, fr-, gr-, pr-, tr-

1. Look at the picture and say the word. Then write the letters of the beginning blend you hear.

1. 

   ____**ib**

4. 

   ____**ass**

2. 

   ____**uck**

5. 

   ____**um**

3. 

   ____**ush**

6. 

   ____**idge**

2. Practice the blends you have learned.

Make at least 10 words with these beginnings and endings.

br	ank	________	________
cr	op	________	________
dr	ush	________	________
fr	ess	________	________
gr	ag	________	________
pr	ip	________	________
tr			

3. Fill in the blends to make sense in the sentences.

cr tr br gr dr fr pr

1. Fred ____ipped and fell down on the bridge.
2. My friend Greg ____ings me some ____esh bread.
3. Brad's ____uck crashed into a van.
4. That brat likes to play tricks and ____anks on us.
5. Fran fixes ____ab and grilled fish for dinner.
6. Put the ____im on the edge of that dress.

   Then ____ess the dress. It will be ready for the ____om.
7. She ____ops the cup. It has a ____ack in it.
8. Never take any ____ugs when you are drinking.
9. Frank ____abs the trunk and ____ags it to the truck.

Practice 17: st-

1. **st-**

tack	top
stack	stop

2. What does the word begin with?

s **t** **st**

1. ______ 9. ______
2. ______ 10. ______
3. ______ 11. ______
4. ______ 12. ______
5. ______ 13. ______
6. ______ 14. ______
7. ______ 15. ______
8. ______ 16. ______

3. Write the letters and say the word.

st		st	
______	ab	______	ill
______	ack	______	itch
______	aff	______	ing
St	an	______	ink
______	em	______	ock
______	ep	______	op
______	ick	______	uck
______	iff	______	uff

4. Read the words.

stop	stick
stock	stink
staff	stitch
stab	still
Stan	stuff
stack	stuck
sting	step
stiff	stem

5. Write the word you hear.

1. ______ 9. ______
2. ______ 10. ______
3. ______ 11. ______
4. ______ 12. ______
5. ______ 13. ______
6. ______ 14. ______
7. ______ 15. ______
8. ______ 16. ______

6. Read the sentences.
Review words: stand, word

1. Ed stops on the steps. He stands still.
2. Our school staff works hard.
3. My finger stings where I cut it.
4. Dad stuffs the hen with bread stuffing.
5. The fudge stuck to the pan.
6. His words were like a stab in the back.
7. I still do not have that hat in stock.
8. The tent is stiff and hard to stitch.
9. Stan keeps the buds fresh.
   He puts the stems in water.
10. Dad stacked the dishes in the sink.
11. Stop playing with that stick.
12. The apples are rotting. They stink!
13. Stan's car was stuck in the ditch.
14. Dr. Bell will stitch up the cut.
15. If your legs get stiff, stand up.

Practice 18: sp-

1. **sp-**

pill	pot
spill	spot

2. What does the word begin with?

s **p** **sp**

1. ______	9. ______
2. ______	10. ______
3. ______	11. ______
4. ______	12. ______
5. ______	13. ______
6. ______	14. ______
7. ______	15. ______
8. ______	16. ______

3. Write the letters and say the word.

sp	**sp**
______ an	______ ill
______ at	______ in
______ ank	______ it
______ eck	______ ot
______ ed	______ un
______ ell	

4. Read the words.

spot	spank
spell	spun
sped	spit
speck	spill
spat	spin
span	

5. Write the word you hear.

1. ____________	7. ____________
2. ____________	8. ____________
3. ____________	9. ____________
4. ____________	10. ____________
5. ____________	11. ____________
6. ____________	

6. Read the sentences.

1. Can you spell this word?
2. The hot sun makes my head spin.
3. The baby spit up on her dress.
4. Stan spilled his drink on the rug. He mopped up the spot with a rag.
5. Jim sped up the street in his car.
6. We made up after our little spat.
7. There are black specks on this paper.
8. Tim spits at his sister. Mom will spank him.
9. The baby spun the top.
10. Jack has spots on his glasses.
11. We are having a hot spell.
12. Pam spilled ink on her skirt.
13. What is the span of that bridge?
14. The hot rod spins down the track.

Practice 19: sn-

1. **sn-**

nag	nip
snag	snip

2. What does the word begin with?

s **n** **sn**

1. ______
2. ______
3. ______
4. ______
5. ______
6. ______
7. ______
8. ______
9. ______
10. ______
11. ______
12. ______
13. ______
14. ______
15. ______
16. ______

3. Write the letters and say the word.

sn		sn	
______	ack	______	itch
______	ag	______	ob
______	ap	______	uff
______	iff	______	ug
______	ip		

4. Read the words.

snip	snag
snitch	snap
sniff	snack
snug	snob
snuff	

5. Write the word you hear.

1. ____________________
2. ____________________
3. ____________________
4. ____________________
5. ____________________
6. ____________________
7. ____________________
8. ____________________
9. ____________________

6. Read the sentences.
Review word: snake

1. I snap my fingers and my pup runs up.
2. That snob thinks he is better than we are.
3. I snagged my dress getting in the car.
4. Our children are snug in their beds.
5. A snake is in the tent.
6. That pup is sniffing at our ham.
7. The baby has a snack after his nap.
8. I want some snaps to put on my skirt.
9. Kim snips an ad out of the paper.
10. That shirt is a little snug for me.
11. The man is sniffing. He must be sick.
12. We have apples and nuts for snacks.
13. The snob thinks his car is best.
14. The pup snapped at my friend's leg.
15. Tom snitches his sister's apple.
16. Our plans hit a snag.

Practice 20: sm-

1. sm-

mash mock

smash smock

2. What does the word begin with?

s **m** **sm**

1. _______ 9. _______
2. _______ 10. _______
3. _______ 11. _______
4. _______ 12. _______
5. _______ 13. _______
6. _______ 14. _______
7. _______ 15. _______
8. _______ 16. _______

3. Write the letters and say the word.

sm

_______ ack

_______ ash

_______ ell

_______ ock

_______ udge

4. Read the words.

smell

smock

smudge

smash

smack

5. Write the word you hear.

1. ______________________________
2. ______________________________
3. ______________________________
4. ______________________________
5. ______________________________

6. Read the sentences.
New words: smog, smart
Review words: cover, Smith

1. The car smashed into a bridge.
2. Cars and factories make smog.
3. It is smart to get a good little car.
4. Ed gives Jill a big smack on the lips. "You smell good to me," he says.
5. Ron has a smudge on his chin.
6. Jan puts on a smock to cover her dress.
7. The stuff in that trash can smells bad.
8. Smog covers the city.
9. Pat smacked the bug with the paper.
10. Dan Smith thinks he is so smart!
11. There is a smudge on this paper.
12. He smashed the trash down in the can.
13. It is smart to get rid of smog.
14. We can smell the dinner. We smack our lips.

Practice 21: sk-, sc-

1. **sk-** **sc-**

sk-	sc-
kid	cab
skid	scab

2. What does the word begin with?

s k sk	**s c sc**
1. _______	1. _______
2. _______	2. _______
3. _______	3. _______
4. _______	4. _______
5. _______	5. _______
6. _______	6. _______
7. _______	7. _______
8. _______	8. _______

3. Write the letters and say the word.

sk	**sc**
_______ etch	_______ ab
_______ id	_______ an
_______ ill	Sc ott
_______ im	_______ uff
_______ in	
_______ ip	
_______ it	
_______ unk	

4. Read the words.

skim	Scott
skit	scuff
skill	scan
skin	scab
skip	
skid	
skunk	
sketch	

5. Write the word you hear.

1. _______________	7. _______________
2. _______________	8. _______________
3. _______________	9. _______________
4. _______________	10. _______________
5. _______________	11. _______________
6. _______________	12. _______________

6. Read the sentences.
New word: scatter
Review words: school, mark

1. That car skids on the wet street.
2. School can help me get job skills.
3. I will sketch a picture of you.
4. The kids scatter when the bell rings.
5. I skip things when I skim a paper.
6. Pam has a new skirt for school.
7. Scott skinned his leg when he fell.
   He has a scab where his skin got cut.
8. The wind scatters her sketches.
9. Jack skipped school to go fishing.
10. Frank scans the paper quickly.
11. When skunks get mad, they smell bad.
12. Do not get any scuff marks on it.
13. Mr. Bell is skilled in fixing cars.
14. Our school will put on a funny skit.
15. There is a scatter rug by my bed.

Practice 22: sw-, tw-

1. **sw-**

well

swell

tw-

win

twin

2. What does the word begin with?

s w sw

1. ______
2. ______
3. ______
4. ______
5. ______
6. ______
7. ______
8. ______

t w tw

1. ______
2. ______
3. ______
4. ______
5. ______
6. ______
7. ______
8. ______

3. Write the letters and say the word.

sw

______ am

______ ell

______ im

______ itch

______ ing

______ ung

tw

______ ig

______ in

______ itch

4. Read the words.

swing

switch

swim

swell

swung

swam

twin

twitch

twig

5. Write the word you hear.

1. ______________________________
2. ______________________________
3. ______________________________
4. ______________________________
5. ______________________________
6. ______________________________
7. ______________________________
8. ______________________________
9. ______________________________

6. Read the sentences.
New word: twenty
Review words: twelve, watch

1. Ann likes to swim in the river.
2. The twins played on the swings.
3. Jack swung at the bug and hit it.
4. Twelve men and twenty women were there.
5. Twigs and little sticks burn quickly.
6. The twins like to switch jobs.
7. Twenty of my friends will come.
8. A bug bit Ed, and his hand swelled up.
9. There are twelve eggs in that box.
10. A school of fish swam by.
11. Dad swings the baby up in his arms.
12. "Your new dress looks swell," Ed said.
13. Switch on the TV. I want to watch it.
14. Hank swung the bat and got a hit!
15. The pup's leg is twitching.
16. The slacks sell for twenty dollars.

Practice 23: Review of Beginning s and w Blends: st-, sp-, sn-, sm-, sk-, sc-, sw-, tw-

1. Look at the picture and say the word. Then write the letters of the beginning blend you hear.

1. 

____**unk**

4. 

____**ing**

2. **20**

____**enty**

5. 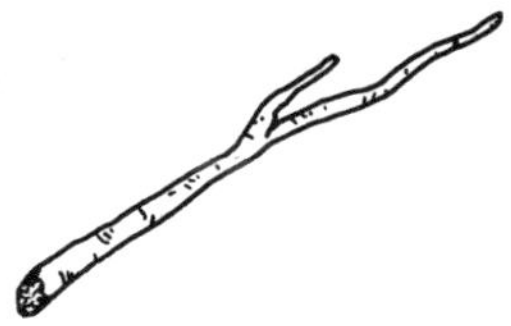

____**ick**

3. 

____**ep**

6. 

____**ake**

2. Practice the blends you have learned.
Make at least 14 words with these beginnings and endings.

st	ell	________	________
sp	itch	________	________
sn	in	________	________
sm	ack	________	________
sc	uff	________	________
sk	ock	________	________
sw		________	________
tw		________	________

3. Fill in the blends to make sense in the sentences.

sc sk sm sn st sw tw sp

1. Scott is still ____imming. He never ____ops!
2. Stan ____ills the ink. It makes a ____ot on the rug.
3. That ____unk stinks! It ____ells bad!
4. My ____in brother ____ings the bat for a hit.
5. She got a ____udge on her sketch.
6. His car ____ids into the ditch. It gets ____uck.
7. ____itch the ____aps on the smock.
8. Scott skips the words he cannot ____ell.
9. Stan ____ins his arm. He gets a ____ab on it.
10. He has a big stock of apples for ____acks.

Practice 24: str-

1. str-

rap
trap
strap

2. What does the word begin with?

st **tr** **str**

1. ______	9. ______
2. ______	10. ______
3. ______	11. ______
4. ______	12. ______
5. ______	13. ______
6. ______	14. ______
7. ______	15. ______
8. ______	16. ______

3. Write the letters and say the word.

str	**str**
______ ap	______ uck
______ ess	______ um
______ etch	______ ut
______ ing	______ ung
______ ip	

4. Read the words.

strip	strum
string	strap
strut	stretch
struck	stress
strung	

5. Write the word you hear.

1. ______________________
2. ______________________
3. ______________________
4. ______________________
5. ______________________
6. ______________________
7. ______________________
8. ______________________
9. ______________________

6. Read the sentences.
Review word: street

1. Ann stretches her arms and legs.
2. The kids are playing in the street.
3. This string is no good.
4. The red van struck our car.
5. Dan is under stress on his job.
6. Molly straps the pack on her back.
7. I got up when the clock struck six.
8. I have strung up the net. Let us play.
9. Ed cut the paper into thin strips.
10. This strap will stretch.
11. Tom struck a match to burn the twigs.
12. Bill strums the strings and sings.
13. The winner struts up and down.
14. We stress good spelling in our class.
15. After work, I stretch out on my bed.
16. The girl's bag hung by a strap.

Practice 25: scr-, spr-, spl-, squ-

1.

scr-	**spr-**
cram	ring
scram	spring
spl-	**squ-**
platter	quint
splatter	squint

2. What does the word begin with?

scr spl spr squ

1. ______
2. ______
3. ______
4. ______
5. ______
6. ______
7. ______
8. ______

3. Write the letters and say the word.

scr

______ am
______ ap
______ atch
______ ub

spr

______ ang
______ ing

squ

______ int

spl

______ ash
______ it

4. Read the words.

scrub
scram
scratch
scrap

spring
sprang

split
splash

squint

5. Write the word you hear.

1. ____________________
2. ____________________
3. ____________________
4. ____________________
5. ____________________
6. ____________________
7. ____________________
8. ____________________
9. ____________________

6. Read the sentences.

1. Scrub your hands after you work.
2. I write the number on some scrap paper.
3. Split the apple in two. Give me half.
4. The park is pretty in the spring.
5. I cannot stand cats that scratch me. I tell them to scram!
6. The passing car splashed mud on me.
7. This bed has no springs.
8. Mom fed the pup some scraps of ham.
9. The sun makes me squint.
10. We get our water from that spring.
11. Fred scrubs the kitchen with a mop.
12. Ben sprang out of bed at six o'clock.
13. Ted got into the tub with a splash.
14. Sam and Jill will split up.
15. Pam scratched her hand fixing her car.

Practice 26: shr-, thr-

1. **shr-** **thr-**

rug	rush
shrug	thrush

2. What does the word begin with?

sh r shr	**th r thr**
1. _______	1. _______
2. _______	2. _______
3. _______	3. _______
4. _______	4. _______
5. _______	5. _______
6. _______	6. _______
7. _______	7. _______
8. _______	8. _______

3. Write the letters and say the word.

shr	**thr**
_______ ed	_______ ill
_______ ill	_______ ush
_______ ink	
_______ ub	
_______ ug	

4. Read the words.

shrug	thrush
shrub	thrill
shrink	
shrill	
shred	

5. Write the word you hear.

1. ________________________________
2. ________________________________
3. ________________________________
4. ________________________________
5. ________________________________
6. ________________________________
7. ________________________________

6. Read the sentences.
New word: thread
Review words: three, nurse

1. We have shrubs in our garden.
2. It thrills me when you sing!
3. I want some pink thread.
4. That shrill whistle makes my head hurt.
5. The nurse works till three o'clock.
6. My shirt is ripped to shreds.
7. I ask Kim if this dress will shrink.
   Kim shrugs and says, "What if it does?"
8. We have three children.
9. I snip a bit of thread from my shirt.
10. A thrush is singing in our garden.
11. I get a thrill out of you, baby!
12. Bob cut the paper into shreds.
13. That dress will shrink if it gets wet.
14. Where do you keep your thread?
15. The pup is digging under the shrubs.

Practice 27: Review of Beginning Blends

1. Look at the picture and say the word. Then write the letters of the beginning blend you hear.

____ab

____and

____elve

____um

____ick

____im

2. Circle all the words that are the same as the first one.

stiff	fits	sniff	stiff	stuff
	stiff	still	stiff	
flap	flop	flap	flat	flip
	flap	lap	flip	
brag	brag	drag	brag	garb
	bag	grab	bran	
spin	spun	spin	pins	spin
	skin	spin	snip	

3. Draw a line from each blend to an ending to make a word.

fl	em	sk	ash
gr	us	bl	ug
pl	esh	sn	ank
st	in	tr	in

4. Use these words to fill in the blanks:

clutch **slim** **bless** **brick**
spit **twigs** **trap** **swell**

1. The school building is made of ________.
2. The slacks make you look ________ and trim!
3. He ________ out the gum and put it in the trash.
4. We want God to ________ this dinner.
5. You have to step on the ________ in this car.
6. We plan to catch rats with this ________.
7. Pick up the ________ and little sticks.
8. The sting is making my arm ________ up.

Practice 28-A: -nt

1. **-nt**

ten plan
tent plant

2. What does the word end with?

n **t** **nt**

1. ______ 9. ______
2. ______ 10. ______
3. ______ 11. ______
4. ______ 12. ______
5. ______ 13. ______
6. ______ 14. ______
7. ______ 15. ______
8. ______ 16. ______

3. Write the letters and say the word.

nt	nt
a ______	le ______
pa ______	re ______
gra ______	se ______
pla ______	te ______
sla ______	ve ______
be ______	we ______
ce ______	spe ______
de ______	

4. Read the words.

lent	rent
vent	tent
bent	pant
sent	slant
dent	grant
went	ant
spent	plant
cent	

5. Write the word you hear.

1. ________ 9. ________
2. ________ 10. ________
3. ________ 11. ________
4. ________ 12. ________
5. ________ 13. ________
6. ________ 14. ________
7. ________ 15. ________
8. ________

6. Read the sentences.
New word: aunt
Review word: twenty

1. Ben sent the rent check to Mrs. Bell.
2. I spent twenty dollars for the pants.
3. Aunt Molly sent me a gift.
4. We have a vent in our kitchen.
5. They are living in a tent by the river.
6. Twenty of us went to work at the plant.
7. I asked Aunt Peg to grant me one wish.
8. Dan was panting as he ran up the steps.
9. Scott lent her twenty cents.
10. My aunt and uncle plant a garden.
11. There are ants and bugs in the tent.
12. That picture is hung on a slant.
13. Ned and Jan are renting a truck.
14. Aunt Pam bent down to look at a plant.
15. Ed got a dent in his car.
He spent twenty dollars to fix it.

Practice 28-B: -nt

1. **nt-**

tin	stun
tint	stunt

2. *-nt* and *-nd* words

be**nd**	be**nt**
le**nd**	le**nt**
se**nd**	se**nt**
spe**nd**	spe**nt**

3. Write the letters and say the word.

nt	nt
hi ______	hu ______
li ______	pu ______
mi ______	blu ______
ti ______	gru ______
fli ______	stu ______
pri ______	
spli ______	
squi ______	

4. Read the words.

blunt	lint
hunt	flint
stunt	splint
punt	mint
grunt	squint
tint	hint
print	

5. Write the word you hear.

1. ______________	8. ______________
2. ______________	9. ______________
3. ______________	10. ______________
4. ______________	11. ______________
5. ______________	12. ______________
6. ______________	13. ______________
7. ______________	

6. Read the sentences.
New word: yard
Review words: another, want

1. I drop hints for the gift I want.
2. Print your name in this blank.
3. This dress is covered with lint.
4. The stunt man fell and hurt his arm.
   We put a splint on his arm.
5. He can kick a punt very well.
6. Birds hunt for bugs in our yard.
7. That one has a blunt edge.
8. Ann tints her hair red.
9. We will plant mint in our back yard.
10. What do you want? Give me a hint.
11. Do not squint. Put on your glasses.
12. I want some help in our yard.
13. When you shop, get me some flints.
14. The pig grunts as it hunts for nuts.
15. I want another print of this picture.

Practice 29-A: -nd

1. **-nd**

men	ten
mend	tend

2. What does the word end with?

n **d** **nd**

1. ______ 9. ______
2. ______ 10. ______
3. ______ 11. ______
4. ______ 12. ______
5. ______ 13. ______
6. ______ 14. ______
7. ______ 15. ______
8. ______ 16. ______

3. Write the letters and say the word.

nd	nd
e ______	bo ______
be ______	fo ______
le ______	po ______
me ______	blo ______
se ______	
te ______	
ble ______	
spe ______	

4. Read the words.

pond	send
bond	lend
blond	bend
fond	spend
mend	tend
end	blend

5. Write the word you hear.

1. ____________ 7. ____________
2. ____________ 8. ____________
3. ____________ 9. ____________
4. ____________ 10. ____________
5. ____________ 11. ____________
6. ____________ 12. ____________

6. Read the sentences.
New words: tender, money
Review word: friend

1. Bob lends me money for the trip.
2. Her hands are red and tender.
3. I send letters to friends I am fond of.
4. Mom bends down to pick up her son.
5. Jim tends to spend a lot of money. We must put an end to his spending.
6. Pam is mending a rip in the dress.
7. Dad sends Ed to the pond to catch fish.
8. Blend the eggs into the mix.
9. Never send money in a letter.
10. My husband says many tender things.
11. My friend Ben is very fond of blonds.
12. Ned will be happy to tend the kids.
13. They run to the end of the street.
14. Jan spends money on stocks and bonds.
15. Scott gives a tender kiss to Jan.

Practice 29-B: -nd

1. **-nd**

ban	win
band	wind

2. What does the word end with?

nd **nt**

1. _______	9. _______
2. _______	10. _______
3. _______	11. _______
4. _______	12. _______
5. _______	13. _______
6. _______	14. _______
7. _______	15. _______
8. _______	16. _______

3. Write the letters and say the word.

nd	**nd**
a _______	gra _______
ba _______	sa _______
ha _______	sta _______
la _______	stra _______
gla _______	wi _______
bra _______	fu _______

4. Read the words.

wind	band
fund	gland
sand	hand
land	brand
and	grand
stand	strand

5. Write the word you hear.

1. _______________	7. _______________
2. _______________	8. _______________
3. _______________	9. _______________
4. _______________	10. _______________
5. _______________	11. _______________
6. _______________	12. _______________

6. Read the sentences.
New words: grandmother, Andy

1. Jill has a wedding band on her hand.
2. This land is covered with sand.
3. Wind and water can chap your hands.
4. I can stand up, but I cannot bend down.
5. We had a grand fishing trip with Andy.
6. The school is running out of funds.
7. They cannot land in this bad wind.
8. Grandmother's school is still standing.
9. The kids are playing in the sand.
10. Andy plays drums in the band.
11. Grandmother sends us jams and jellies.
12. My car got stuck, and I was stranded.
13. I cannot stand this wind!
14. Grandmother is mending Andy's shirt.
15. Hand me that can of corn.
    I like that brand.

Practice 30: -nch

1. **-nch**

pin	bran
pinch	branch

2. What does the word end with?

n **ch** **nch**

1. _______ 9. _______
2. _______ 10. _______
3. _______ 11. _______
4. _______ 12. _______
5. _______ 13. _______
6. _______ 14. _______
7. _______ 15. _______
8. _______ 16. _______

3. Write the letters and say the word.

nch	nch
ra _______	pi _______
bra _______	bu _______
be _______	hu _______
que _______	lu _______
dre _______	mu _______
tre _______	pu _______
i _______	cru _______

4. Read the words.

drench	crunch
trench	bunch
quench	munch
bench	pinch
hunch	inch
punch	ranch
lunch	branch

5. Write the word you hear.

1. _______________ 8. _______________
2. _______________ 9. _______________
3. _______________ 10. _______________
4. _______________ 11. _______________
5. _______________ 12. _______________
6. _______________ 13. _______________
7. _______________ 14. _______________

6. Read the sentences.
Review word: valley

1. We have a big ranch in the valley.
2. I have a hunch that Jack wants to go.
3. I fell in the water and got drenched.
4. When I pinched her arm, she yelled.
5. A bunch of us work at a branch office.
6. Take up the pants an inch and a half.
7. Andy made some punch for our lunch.
8. The apple crunched when he bit into it.
9. The batter went back to the bench.
10. A bunch of branches fell in our yard.
11. That car missed me by inches.
12. I have a hunch I am going to win.
13. The ranch hands stop work for lunch.
14. I picked up an apple to munch on.
15. Stan punched in at seven o'clock.
16. She digs a trench in her yard.

Practice 31: -nce, -nse, -nge

1.

-nce	-nse	-nge
sin	ten	twin
since	tense	twinge

2. What does the word end with?

n nce	n nge
1. _______	1. _______
2. _______	2. _______
3. _______	3. _______
4. _______	4. _______
5. _______	5. _______
6. _______	6. _______
7. _______	7. _______
8. _______	8. _______

3. Write the letters and say the word.

nce	nse	nge
da _______	se _______	hi _______
cha _______	te _______	fri _______
gla _______	ri _______	plu _______
tra _______		
fe _______		
si _______		
pri _______		

4. Read the words.

fence	rinse	plunge
prince	sense	fringe
since	tense	hinge
chance		
trance		
dance		
glance		

5. Write the word you hear.

1. _______________	8. _______________
2. _______________	9. _______________
3. _______________	10. _______________
4. _______________	11. _______________
5. _______________	12. _______________
6. _______________	13. _______________
7. _______________	

6. Read the sentences.
New word: once

1. Jan puts up a fence to keep her pup in.
2. He wants a chance to dance with her.
3. I glance at the clock and rush to work.
4. Will you fix the hinge on my trunk?
5. I went to a dance there once.
6. It is good sense not to take chances.
7. I rinse my hair with hot water.
8. Jill has danced since she was six.
9. They plunge into the river to swim.
10. That does not make sense to me.
11. I danced with a prince once.
12. She looks like she is in a trance.
13. A red fence hid the trash cans.
14. I get tense when the kids yell.
15. Once I had a chance to marry Pat.
16. Ann put a fringe on her skirt.

Practice 32: -ng

1. **-ng**

win	ban
wing	bang

2. What does the word end with?

n **g** **ng**

1. _______	9. _______
2. _______	10. _______
3. _______	11. _______
4. _______	12. _______
5. _______	13. _______
6. _______	14. _______
7. _______	15. _______
8. _______	16. _______

3. Write the letters and say the word.

ng	**ng**
ba _______	bri _______
ga _______	sli _______
ha _______	sti _______
ki _______	swi _______
ri _______	spri _______
si _______	stri _______
wi _______	hu _______
thi _______	lu _______

4. Read the words.

ring	sing
thing	swing
sting	spring
king	lung
sling	hung
bring	gang
wing	bang
string	hang

5. Write the word you hear.

1. ____________	9. ____________
2. ____________	10. ____________
3. ____________	11. ____________
4. ____________	12. ____________
5. ____________	13. ____________
6. ____________	14. ____________
7. ____________	15. ____________
8. ____________	16. ____________

6. Read the sentences.
New word: something
Review word: finger

1. The ring does not fit my finger.
2. Bob hangs out with a street gang.
3. I think something is stinging me!
4. Andy hung up the telephone.
5. The doctor put Ed's arm in a sling.
6. Bring me something to put on.
7. He has a big string of fish.
8. Jack will sing at the spring dance.
9. The king asks for something to drink.
10. Scott bangs on his drums.
11. They fix the wing of the jet.
12. We stop work when the bell rings.
13. A gang of kids plays on the swings.
14. Smog hurts your lungs.
15. The picture is hanging by a string.
16. The King Family sings on TV.

1. **-nk**

wing bang

wink bank

2. What does the word end with?

n **k** **nk**

1. ______ 9. ______
2. ______ 10. ______
3. ______ 11. ______
4. ______ 12. ______
5. ______ 13. ______
6. ______ 14. ______
7. ______ 15. ______
8. ______ 16. ______

3. Write the letters and say the word.

nk	nk
ba ______	wi ______
ra ______	thi ______
ta ______	dri ______
ya ______	sti ______
li ______	shri ______
pi ______	bu ______
ri ______	du ______
si ______	ju ______

4. Read the words.

wink	drink
pink	dunk
stink	junk
think	bunk
link	yank
sink	rank
rink	bank
shrink	tank

5. Write the word you hear.

1. ______ 9. ______
2. ______ 10. ______
3. ______ 11. ______
4. ______ 12. ______
5. ______ 13. ______
6. ______ 14. ______
7. ______ 15. ______
8. ______ 16. ______

6. Read the sentences.
Review words: past, thank, Frank, Hank

1. I think we have money in the bank.
2. Frank winks at Peg.
3. Do not dunk that pot in water.
4. He ranks number one in my class.
5. Thanks for giving us the bunk beds.
6. I think that is funny!
7. Put this junk in the trash can.
8. Hank drinks a glass of water.
9. The fish sinks in the tank.
10. She fell down at the rink.
11. Fill out this pink slip.
12. My pants will shrink if they get wet.
13. This trash stinks!
14. He yanks the paper out of my hands.
15. Grandmother is my link with the past.
16. Thanks for filling up the gas tank.

Practice 34: More Practice with -ng and -nk

1. **-ng** wing bang sung

   -nk wink bank sunk

What does the word end with?

ng nk

1. _______	12. _______	23. _______
2. _______	13. _______	24. _______
3. _______	14. _______	25. _______
4. _______	15. _______	26. _______
5. _______	16. _______	27. _______
6. _______	17. _______	28. _______
7. _______	18. _______	29. _______
8. _______	19. _______	30. _______
9. _______	20. _______	31. _______
10. _______	21. _______	32. _______
11. _______	22. _______	

2. The **present** is used for things that happen again and again.

I **sing** every day.

The **past** is used for things that happened and are finished.

I **sang** yesterday.

And some forms of these words are used with **has** or **have**.

I **have sung** since I was ten.

Every day	Yesterday	has/have +
ring	rang	rung
sing	sang	sung
spring	sprang	sprung
drink	drank	drunk
sink	sank	sunk
stink	stank	stunk
shrink	shrank	shrunk

Every day	Yesterday	has/have +
cling	clung	clung
fling	flung	flung
sting	stung	stung
string	strung	strung
swing	swung	swung
hang	hung	hung

3. Use the **past** of the word.

sink 1. The ship _________ yesterday.

hang 2. We _________ up the picture.

fling 3. She _________ her bag down.

ring 4. My telephone _________.

string 5. They _________ up the net.

cling 6. Jimmy _________ to my hand.

drink 7. The baby _________ from a cup.

spring 8. Dad _________ out of bed.

Use the word that goes with **has** or **have**.

ring 1. The bell for class has _________.

hang 2. He has _________ up his shirt.

sing 3. We have _________ at dances.

shrink 4. The dresses have _________.

drink 5. Jim has _________ some water.

stink 6. The trash has _________ for days.

swing 7. Bob has _________ the bat.

sting 8. No bugs have _________ us.

Practice 35: Review of Ending n Blends: -nt, -nd, -nch, -nge, -nce, -nse, -ng, -nk

1. Look at the picture and say the word. Then write the letters of the ending blend you hear.

1. 

wi_____

4. 

ki_____

2. 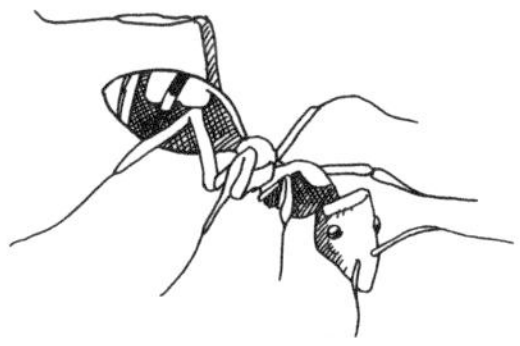

a_____

5. 

ba_____

3. 

be_____

6. 

fe_____

2. Practice the blends you have learned.

Make at least 12 words with these beginnings and endings.

pu	nd	__________	__________
wi	nt	__________	__________
ra	nch	__________	__________
be	ng	__________	__________
hu	nk	__________	__________
		__________	__________
		__________	__________

3. Fill in the blends to make sense in the sentences.

nk nt nch nce nse nd nge ng

1. Andy's aunt went out to lu_____ and spent twenty dollars.
2. This bunch of numbers makes no se_____ to me.
3. She has tended the pla_____s since spring.
4. The ga_____ hangs out at the ranch every cha_____ they get.
5. Tha_____ you for bringing me some punch to dri_____.
6. The blo_____ girl can sing and da_____.
7. I think I will take a plu_____ in the river.
8. Give me a ha_____. I will pitch this tent on the sa_____.
9. The pi_____ pants are covered with li_____.
10. I spe_____ the funds. But the bank can le_____ me money.

Practice 36: -mp

1. -mp

dam	bum
damp	bump

2. What does the word end with?

m **p** **mp**

1. _______	9. _______
2. _______	10. _______
3. _______	11. _______
4. _______	12. _______
5. _______	13. _______
6. _______	14. _______
7. _______	15. _______
8. _______	16. _______

3. Write the letters and say the word.

mp	mp
ca _______	li _______
da _______	shri _______
la _______	bu _______
ra _______	du _______
cha _______	lu _______
cra _______	plu _______
tra _______	stu _______
sta _______	pu _______

4. Read the words.

lump	ramp
stump	camp
pump	stamp
dump	lamp
bump	tramp
plump	damp
shrimp	champ
limp	cramp

5. Write the word you hear.

1. _______________	9. _______________
2. _______________	10. _______________
3. _______________	11. _______________
4. _______________	12. _______________
5. _______________	13. _______________
6. _______________	14. _______________
7. _______________	15. _______________
8. _______________	16. _______________

6. Read the sentences.
Review word: jump

1. At camp, we get water from a pump.
2. Take that lamp to the dump.
3. Don has a cramp in his leg.
   It makes him limp.
4. This letter has no stamp on it.
5. Tom bumped his head and got a big lump.
6. Ned dumps the shrimp in a pan.
7. I jump in my car and head for our camp.
8. I will dig up the stump in my yard.
9. We hung up the damp shirts.
10. There are bumps in our street.
11. Jim is plump. He wants to be thin.
12. Rub the lamp with a damp rag.
13. We look like tramps.
14. He is the winner and still champ!
15. The car got on the ramp.

Practice 37-A: -st

1. -st

fat	pet
fast	pest

2. What does the word end with?

s t st

1. ______ 9. ______
2. ______ 10. ______
3. ______ 11. ______
4. ______ 12. ______
5. ______ 13. ______
6. ______ 14. ______
7. ______ 15. ______
8. ______ 16. ______

3. Write the letters and say the word.

st	st
ca ______	gue ______
fa ______	ne ______
la ______	pe ______
ma ______	re ______
pa ______	te ______
va ______	ve ______
bla ______	we ______
be ______	che ______

4. Read the words.

test	pest
nest	last
chest	cast
guest	mast
vest	blast
best	fast
west	vast
rest	past

5. Write the word you hear.

1. ______ 9. ______
2. ______ 10. ______
3. ______ 11. ______
4. ______ 12. ______
5. ______ 13. ______
6. ______ 14. ______
7. ______ 15. ______
8. ______ 16. ______

6. Read the sentences.
New word: day
Review word: left

1. The car is headed west, going fast.
2. The bird builds a nest out of twigs.
3. Do your best on the rest of the test.
4. Dr. King put my leg in a cast.
5. The last guest left at half past ten.
6. This brand of pest killer is the best.
7. Ted ran fast. He ran past the shop.
8. Dan's ranch out west is vast.
9. He will test that car last.
10. Put this red vest in the chest.
11. The ship has a big mast.
12. No one was hurt in the blast.
13. I will fast for the rest of the day.
14. Our guest wants to rest.
15. Ed's chest hurts.
    Dr. Bell tells him to rest.

Practice 37-B: -st

1. **-st**

lit	rut
list	rust

2. Circle the word you hear.

1. nets	nest
2. fits	fist
3. vets	vest
4. ruts	rust
5. bets	best
6. pats	past
7. cats	cast
8. wets	west

3. Write the letters and say the word.

st	st
fi _______	gu _______
li _______	ju _______
mi _______	mu _______
twi _______	ru _______
bu _______	cru _______
du _______	tru _______

4. Read the words.

just	mist
rust	fist
bust	twist
must	list
trust	
gust	
crust	
dust	

5. Write the word you hear.

1. _______________	7. _______________
2. _______________	8. _______________
3. _______________	9. _______________
4. _______________	10. _______________
5. _______________	11. _______________
6. _______________	12. _______________

6. Read the sentences.
New word: off
Review word: sister

1. Cut the crusts off the bread.
2. Dusting is the last job on my list.
3. My car has just started to rust.
4. This thread is twisted.
5. You must have trust in your friends.
6. He punched the bag with his fists.
7. A gust of wind carried off my hat.
8. There is just one test I must take.
9. A thick mist covered the pond.
10. Mr. Hill set up a trust fund.
11. I must make out a shopping list.
12. Put the apple filling into the crust.
13. My sister just dusted off the chest.
14. The pot was covered with rust.
15. You can trust that judge to be just.
16. Dan twisted the lid off the jar.

Practice 38: -sk, -sp

1.

-sk	-sp
mass	gas
mask	gasp

2. What does the word end with?

s k sk	s p sp
1. _______	1. _______
2. _______	2. _______
3. _______	3. _______
4. _______	4. _______
5. _______	5. _______
6. _______	6. _______
7. _______	7. _______
8. _______	8. _______

3. Write the letters and say the word.

sk	sk	sp
a _______	du _______	ga _______
ma _______	hu _______	cla _______
ta _______	tu _______	gra _______
de _______		cri _______
ri _______		
whi _______		
bri _______		

4. Read the words.

desk	brisk	crisp
husk	whisk	gasp
tusk	mask	grasp
dusk	ask	clasp
risk	task	

5. Write the word you hear.

1. _______________	8. _______________
2. _______________	9. _______________
3. _______________	10. _______________
4. _______________	11. _______________
5. _______________	12. _______________
6. _______________	13. _______________
7. _______________	14. _______________

6. Read the sentences.
New word: boy
Review words: corn, basket

1. The farmer worked until dusk.
2. Tom asked if there was a big risk.
3. The boys and girls sit at their desks.
4. The cab whisked him off to his office.
5. This is a list of tasks for you to do.
6. My little boy has a funny mask.
7. I was gasping after I ran up the steps.
8. I step out into the crisp, brisk wind.
9. Her task was to cut the grass.
10. Put the corn husks into the basket.
11. You take a risk when you bet money.
12. That boy put on his catcher's mask.
13. This bread has a crisp crust.
14. She grasped my hand and clasped it.
15. Ed asks if he can work at her desk.
16. The pig has two big tusks.

Practice 39: -lt, -ld

1.

-lt	-ld
bell	well
belt	weld

2. What does the word end with?

l t lt	l d ld
1. _______	1. _______
2. _______	2. _______
3. _______	3. _______
4. _______	4. _______
5. _______	5. _______
6. _______	6. _______
7. _______	7. _______
8. _______	8. _______

3. Write the letters and say the word.

lt	ld
be _______	he _______
fe _______	we _______
me _______	bui _______
bui _______	
gui _______	
qui _______	
ti _______	
wi _______	

4. Read the words.

tilt	build
built	weld
wilt	held
quilt	
guilt	
melt	
belt	
felt	

5. Write the word you hear.

1. _______________	7. _______________
2. _______________	8. _______________
3. _______________	9. _______________
4. _______________	10. _______________
5. _______________	11. _______________
6. _______________	

6. Read the sentences.
New words: over, together
Review word: butter

1. Grandmother is making a quilt.
2. The belt held up his pants.
3. Melt the fat in the pan.
4. Dad built a desk. It held together.
5. The robber cannot get rid of his guilt.
6. My plant is wilting.
7. We worked together to build the school.
8. When he tilted it back, it fell over.
9. Andy welds two things together.
10. Dan plans to build six buildings.
11. The belt felt good when I put it on.
12. Put the quilt over the bed.
13. Mom held her baby. The baby felt wet.
14. They built a bridge over the river.
15. She felt guilt over what happened.
16. Put the melted butter on the corn.

Practice 40: -lk, -lf, -lp, -lm

1.

-lk	-lf	-lp	-lm
mill	shell	hell	fill
milk	shelf	help	film

2. What does the word end with?

l k lk

1. _______
2. _______
3. _______
4. _______
5. _______
6. _______
7. _______
8. _______

l f lf

1. _______
2. _______
3. _______
4. _______
5. _______
6. _______
7. _______
8. _______

3. Write the letters and say the word.

lk

e _______
mi _______
si _______

lf

e _______
se _______
she _______
go _______
gu _______

lp

sca _______
he _______
ye _______
gu _______

lm

e _______
fi _______

4. Read the words.

silk	help	golf	film
milk	yelp	self	elm
elk	gulp	shelf	
	scalp	elf	
		gulf	

5. Write the word you hear.

1. _______________
2. _______________
3. _______________
4. _______________
5. _______________
6. _______________
7. _______________
8. _______________
9. _______________
10. _______________
11. _______________
12. _______________
13. _______________
14. _______________

6. Read the sentences.

1. I am making a silk dress for myself.
2. The ship headed into the gulf.
3. I take colored pictures with this film.
4. Pam gulped down a glass of milk.
5. Stan plays golf by himself.
6. We helped Dad cut down the big elm.
7. Scott will put up the shelf himself.
8. The baby wants to do things for herself.
9. My uncle is in the Elks Club.
10. Frank helps himself to some milk.
11. The little elf had fun playing tricks.
12. The silk thread is on the top shelf.
13. She helped me when I hurt myself.
14. The class is watching a good film.
15. The pup yelps when it is hurt.
16. This hat makes my scalp itch!

Practice 41: -ft

1. **-ft**

rat	sit
raft	sift

2. What does the word end with?

f **t** **ft**

1. _______	9. _______
2. _______	10. _______
3. _______	11. _______
4. _______	12. _______
5. _______	13. _______
6. _______	14. _______
7. _______	15. _______
8. _______	16. _______

3. Write the letters and say the word.

ft	ft
ra _______	li _______
sha _______	si _______
cra _______	shi _______
dra _______	thri _______
le _______	dri _______
the _______	swi _______
gi _______	

4. Read the words.

sift	craft
drift	raft
shift	shaft
swift	draft
thrift	theft
gift	left
lift	

5. Write the word you hear.

1. ____________	8. ____________
2. ____________	9. ____________
3. ____________	10. ____________
4. ____________	11. ____________
5. ____________	12. ____________
6. ____________	13. ____________
7. ____________	

6. Read the sentences.
New word: after

1. The raft drifted down the river.
2. Sift it to get the lumps out.
3. That draft is giving me a chill.
4. Kim cannot lift her arm.
5. I stress thrift. I do not spend much.
6. I had no money left to pay for the gift.
7. That man is wanted for car theft.
8. We work the last shift at the factory.
9. We asked him for a lift into the city.
10. After one, I left for my crafts class.
11. My aunt works with gifted children.
12. Jack drifted from one job to another.
13. Jill's car has a stick shift.
14. Ned is a swift runner.
15. He tripped and fell down the dark shaft.
16. Ron is skilled in his craft.

Practice 42: -pt, -ct, -xt

1.

-pt	-ct	-xt
wet	fat	net
wept	fact	next

2. What does the word end with?

p t pt	c t ct
1. _______	1. _______
2. _______	2. _______
3. _______	3. _______
4. _______	4. _______
5. _______	5. _______
6. _______	6. _______
7. _______	7. _______
8. _______	8. _______

3. Write the letters and say the word.

pt	ct	xt
a _______	a _______	ne _______
ke _______	fa _______	te _______
we _______	stri _______	
cre _______		
sle _______		
swe _______		
scri _______		

4. Read the words.

crept	fact	text
wept	act	next
slept	strict	
kept		
swept		
script		
apt		

5. Write the word you hear.

1. _______	7. _______
2. _______	8. _______
3. _______	9. _______
4. _______	10. _______
5. _______	11. _______
6. _______	12. _______

6. Read the sentences.
New words: study, adopt, expect
Review words: factory, doctor

1. The judge will study the facts.
2. We kept expecting something to happen.
3. We want to adopt our next kid.
4. Dad acted fast. He swept up the glass.
5. The doctor has not slept much for days.
6. The crab crept over the sand.
7. I expect Ed will act in our next play. He is studying the script for it.
8. We expect to adopt her plan.
9. This is the next text we will study.
10. Pam felt so sad that she wept.
11. The baby crept into the kitchen.
12. He is apt to be strict if they act up.
13. The pup slept next to my bed.
14. This text gives facts on adopted kids.
15. John kept his job at the factory.

Practice 43: Review of Other Ending Blends: -mp, -st, -sk, -sp, -lk, -lf, -lm, -lp, -ct, -pt, -xt

1. Look at the picture and say the word. Then write the letters of the ending blend you hear.

1. 

mi_____

2. 

pu_____

3. 

de_____

4. 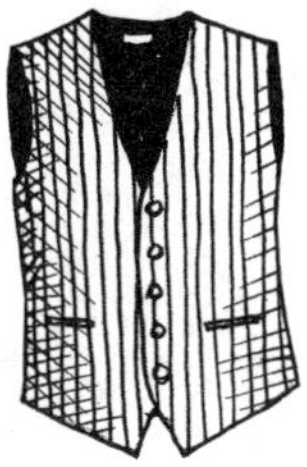

ve_____

5. 

gi_____

6. 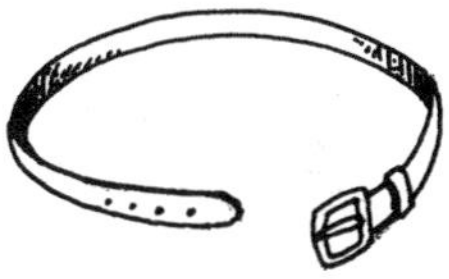

be_____

2. Circle all the words that are the same as the first one.

mask	mash	mask	ask	mast	mash	smack	mask
left	lift	felt	left	let	left	slept	left
past	past	pat	past	pest	pact	past	spat
belt	bet	belt	bell	built	pelt	belt	belt

3. Draw a line from each beginning to an ending blend to make a word.

mi	xt
fa	ft
ne	ct
dra	lk
tu	sp
cri	sk
sta	pt
we	mp

4. Use these words to fill in the blanks:

shelf held film help list guilt west risks

1. The car left the camp, heading ______ on the ramp.
2. Dust the ______ and put the lamp on it.
3. My guest asked me if I wanted any ______.
4. The fact was, he felt no ______ for his acts.
5. The champ kept taking ______.
6. Make a ______ of the tasks you do best.
7. I ______ the golf club and lifted it over my head.
8. Andy gasped when he watched the ______.

Practice 44: er

1. er

her	were
fern	jerk

2. Write the letters you hear.

	er			**re**
1. t ______ m			5. ______ d	
2. f ______ t			6. d ______ ss	
3. v ______ b			7. cl ______ k	
4. f ______ n			8. f ______ sh	

3. Write the letters and say the word.

er	**er**
h ______	f ______ n
v ______ b	F ______ n
h ______ d	st ______ n
m ______ ge	v ______ se
j ______ k	n ______ ve
p ______ k	s ______ ve
cl ______ k	sw ______ ve
t ______ m	p ______ ch

4. Read the words.

term	serve
Fern	nerve
fern	swerve
stern	herd
her	perch
perk	verb
jerk	merge
clerk	verse

5. Write the word you hear.

1. ______	9. ______
2. ______	10. ______
3. ______	11. ______
4. ______	12. ______
5. ______	13. ______
6. ______	14. ______
7. ______	15. ______
8. ______	16. ______

6. Read the sentences.
Review words: were, heard, person

1. Fern pays the clerk in the gift shop.
2. I heard him sing the last verse.
3. Make a list of the helping verbs.
4. They serve fresh perch for lunch.
5. Dan is a very strict and stern person.
6. The van stopped with a jerk.
7. I have just one term paper to write.
8. Pam likes ferns best of any plants.
9. The cars merge when they get off the ramp.
10. A hot drink will perk you up.
11. We heard her yelling at the clerk. She has a lot of nerve!
12. The car had to swerve to miss the pup.
13. The bird is resting on the perch.
14. Don will serve two terms in office.
15. We were herding the pigs into the pen.

Practice 45: ir

1. **ir**

bird
girl
skirt

2. Write the letters you hear.

ir **ri**

1. t _______ m
2. d _______ t
3. b _______ ch
4. c _______ b
5. f _______ st
6. g _______ ll
7. t _______ p
8. f _______ m

3. Write the letters and say the word.

ir	ir
f _______	ch _______ p
s _______	d _______ t
st _______	sh _______ t
th _______ d	squ _______ t
wh _______ l	b _______ ch
tw _______ l	b _______ th
f _______ m	f _______ st
squ _______ m	th _______ st

4. Read the words.

firm	sir
squirm	stir
shirt	fir
dirt	birth
squirt	thirst
twirl	first
whirl	third
chirp	birch

5. Write the word you hear.

1. _______________
2. _______________
3. _______________
4. _______________
5. _______________
6. _______________
7. _______________
8. _______________
9. _______________
10. _______________
11. _______________
12. _______________
13. _______________
14. _______________
15. _______________
16. _______________

6. Read the sentences.
New word: tree
Review words: bird, girl, skirt

1. Stir in the eggs first, then the nuts.
2. I got dirt on my shirt and my skirt.
3. We planted birch trees and fir trees.
4. I was thrilled after my son's birth!
5. Dad is firm when he says no.
6. The bird sits in the tree, chirping.
7. You will be first, and I will be third.
8. Pat whirls and twirls when she dances.
9. The little girl sat and squirmed.
10. This drink will quench your thirst.
11. Stir it until it gets firm.
12. The pup got mud and dirt on Ed's shirt.
13. Ed says "Yes, sir" to his father.
14. She is giving birth to her third baby.
15. Fern has on a pink skirt.
16. He squirts water on the thirsty plants.

Practice 46: ur

1. **ur**

hurt

burn

nurse

2. Write the letters you hear.

	ur	ru	
1. c _______ l		5. g _______ ff	
2. d _______ g		6. c _______ sh	
3. b _______ p		7. c _______ b	
4. t _______ n		8. d _______ m	

3. Write the letters and say the word.

ur	ur
f _______	ch _______ n
p _______ r	b _______ p
bl _______	c _______ se
c _______ b	n _______ se
s _______ f	p _______ se
_______ ge	c _______ ve
c _______ l	ch _______ ch
t _______ n	b _______ st

4. Read the words.

curl	curb
churn	purr
turn	fur
purse	blur
curse	church
nurse	burst
surf	curve
urge	burp

5. Write the word you hear.

1. _______________	9. _______________
2. _______________	10. _______________
3. _______________	11. _______________
4. _______________	12. _______________
5. _______________	13. _______________
6. _______________	14. _______________
7. _______________	15. _______________
8. _______________	16. _______________

6. Read the sentences.
New word: wife
Review words: hurt, burn, hurry, curtains, cover

1. My cat purrs when I rub her fur.
2. The car turns and stops at the curb.
3. A man grabbed my purse and hurried off.
4. One person is hurt. His hand is burned.
   A nurse covers the burns on his hand.
5. We sit in the sand and watch the surf.
6. Ted's wife urges him to go to church.
7. She fills up her purse. It is bursting!
8. The farmer's wife churns the butter.
9. Fran is curling her hair.
10. Bill burst into the church office.
11. Jim cursed when he hurt his finger.
12. Hurry up and put up the curtains!
13. My wife is burping our baby.
14. The nurses take turns working.
15. The car went fast on the curve.

Practice 47-A: ar

1. **ar**

car mark
barn park

2. Write the letters you hear.

	ra	ar	
1. d ______ k			5. d ______ g
2. g ______ b			6. sh ______ p
3. b ______ n			7. c ______ sh
4. ______ t			8. p ______ k

3. Write the letters and say the word.

ar	ar
b ______	b ______ k
c ______	d ______ k
f ______	l ______ k
j ______	m ______ k
t ______	M ______ k
sc ______	p ______ k
st ______	sh ______ k
y ______ n	sp ______ k

4. Read the words.

park	yarn
Mark	far
bark	tar
shark	car
dark	jar
spark	star
mark	bar
lark	scar

5. Write the word you hear.

1. ______	9. ______
2. ______	10. ______
3. ______	11. ______
4. ______	12. ______
5. ______	13. ______
6. ______	14. ______
7. ______	15. ______
8. ______	16. ______

6. Read the sentences.
Review words: barn, large, market

1. Get a jar of jelly at the market.
2. The park is not far from the school.
3. Fern gets good marks in her math class.
4. When it is dark, we look at the stars.
5. The pup is barking. He is not far off.
6. I am making a hat out of dark red yarn.
7. Ted put a bar under the box to lift it.
8. Sparks fell from the burning barn.
9. Sharks are very large fish.
10. Kim parked her car next to the barn.
11. The street is covered with black tar.
12. Mark had lunch at the snack bar.
13. He has a large scar on his leg.
14. A lark is singing in the park.
15. That tree has a thick bark.
16. We cannot go very far in this car.

Practice 47-B: ar

1. **ar**

arm

farm

large

2. Write the letters you hear.

	ar	ra	
1. c _______ d		5. sc _______ f	
2. b _______ n		6. b _______ g	
3. _______ sh		7. f _______ m	
4. _______ m		8. g _______ ss	

3. Write the letters and say the word.

ar	ar
c _______ d	h _______ m
gu _______ d	ch _______ m
h _______ d	b _______ ge
l _______ d	l _______ ge
y _______ d	M _______ ge
sc _______ f	ch _______ ge
C _______ l	h _______ p
sn _______ l	sh _______ p

4. Read the words.

sharp	hard
harp	yard
large	guard
charge	lard
barge	card
Marge	charm
snarl	harm
Carl	scarf

5. Write the word you hear.

1. _______________	9. _______________
2. _______________	10. _______________
3. _______________	11. _______________
4. _______________	12. _______________
5. _______________	13. _______________
6. _______________	14. _______________
7. _______________	15. _______________
8. _______________	16. _______________

6. Read the sentences.
Review words: arm, farm, farmer, garden

1. Marge is sending Carl a birthday card.
2. We have a large barnyard on our farm.
3. The pup snarls at me. Will it harm me?
4. My wife plants a garden in the yard.
5. Large barges carry goods up the river.
6. Marge has her good luck charm with her.
7. The farmer works hard on his farm.
8. The bark on that tree is hard.
9. It has a sharp edge. It cuts well.
10. Carl hands the clerk his charge card.
11. Lard is the fat that comes from pigs.
12. A large yarn scarf covers her head.
13. Marge can play the harp.
14. Our garden is six yards by ten yards.
15. Armed guards kept him from harm.
16. There is no charge for parking there.

Practice 47-C: ar

1. **ar**

start

Arthur

march

2. Write the letters you hear.

	ar	**ra**	
1. t ______ p		5. st ______ ch	
2. d ______ t		6. b ______ t	
3. c ______ b		7. t ______ sh	
4. b ______ k		8. p ______ t	

3. Write the letters and say the word.

ar	**ar**
______ t	c ______ ve
c ______ t	st ______ ve
d ______ t	h ______ sh
p ______ t	m ______ sh
t ______ t	______ ch
ch ______ t	m ______ ch
sm ______ t	M ______ ch
st ______ t	st ______ ch

4. Read the words.

part	marsh
chart	harsh
art	March
start	arch
cart	starch
smart	march
dart	carve
tart	starve

5. Write the word you hear.

1. ______	9. ______
2. ______	10. ______
3. ______	11. ______
4. ______	12. ______
5. ______	13. ______
6. ______	14. ______
7. ______	15. ______
8. ______	16. ______

6. Read the sentences.
New words: love, army, Barb
Review words: Arthur, Indian

1. Put a jar of nuts in the shopping cart.
2. The army started marching up the hill.
3. Mom puts starch in Dad's shirts.
4. Arthur loves his arts and crafts class.
5. Barb is making little apple tarts.
6. Water covers a large part of the marsh.
7. Dad started to carve the hen.
8. In March, the army ended the draft.
9. My sister Barb loves to play darts.
10. Arthur is smart. He makes good marks.
11. What part do you act in the play?
12. Her red, chapped hands felt harsh.
13. The head of the army is studying charts.
14. My cat arched her back and hissed.
15. I am starved! When do we have lunch?
16. Barb loves to study Indian art.

Practice 48-A: or

1. **or**

or

for

corn

2. Write the letters you hear.

	or	ro	
1. b ______ n			5. c ______ d
2. t ______ t			6. b ______ th
3. f ______ k			7. ______ b
4. p ______ p			8. t ______ n

3. Write the letters and say the word.

or	or
______	b ______ n
c ______ d	c ______ n
F ______ d	h ______ n
l ______ d	t ______ n
c ______ k	w ______ n
f ______ k	th ______ n
p ______ k	sc ______ n
st ______ k	sw ______ n

4. Read the words.

torn	or
corn	fork
sworn	stork
thorn	cork
horn	pork
scorn	Ford
worn	cord
born	lord

5. Write the word you hear.

1. ______	9. ______
2. ______	10. ______
3. ______	11. ______
4. ______	12. ______
5. ______	13. ______
6. ______	14. ______
7. ______	15. ______
8. ______	16. ______

6. Read the sentences.
New words: morning, new
Review word: York

1. My new Ford has a good horn!
2. The plant is covered with sharp thorns.
3. We were born in New York City.
4. Give me a fork. I will carve this pork.
5. She was sworn into office this morning.
6. Ed trips over the cord of the TV set.
7. Jack has no job, but do not scorn him.
8. He has not worn that shirt. It is torn.
9. Serve corn or bread with the pork.
10. The stork visited them this morning!
    A new baby was born to them!
11. Barb asks, "Lord God, help us!"
12. Put the dishes and forks in the sink.
13. The farmer picks corn every morning.
14. Mark plays his new horn in the band.
15. Put the cork back in to stop it up.

Practice 48-B: or

1. **or**

or

for

short

2. Write the letters you hear.

or **ro**

1. c _______ p	5. t _______ ch
2. f _______ m	6. d _______ p
3. p _______ t	7. f _______ t
4. _______ t	8. p _______ m

3. Write the letters and say the word.

or	**or**
f _______ t	f _______ ce
p _______ t	h _______ se
s _______ t	p _______ ch
sh _______ t	t _______ ch
sp _______ t	sc _______ ch
sn _______ t	f _______ th
f _______ m	n _______ th
st _______ m	f _______ ge

4. Read the words.

torch	forge
porch	force
scorch	sort
storm	port
form	short
horse	snort
north	fort
forth	sport

5. Write the word you hear.

1. _______________	9. _______________
2. _______________	10. _______________
3. _______________	11. _______________
4. _______________	12. _______________
5. _______________	13. _______________
6. _______________	14. _______________
7. _______________	15. _______________
8. _______________	16. _______________

6. Read the sentences.

1. Marge is sitting out on the porch.
2. I fill out the short form to pay taxes. I have to force myself to do it.
3. The army marched north to the fort.
4. What sort of sports do you like best?
5. The horse lifted his head and snorted.
6. Carl forged my name on a check.
7. When it got dark, Ed lit a torch.
8. There is a bad storm north of us.
9. I scorched the shirt. It is burned.
10. Bill is the best cop on the force.
11. I am short, but I play many sports.
12. The ship is coming into port.
13. We are forming a sports club.
14. The women are sorting the letters.
15. The horse went forth into the storm.

Practice 49: Review of Vowels + r: er, ir, ur, ar, or

1. Look at the picture and say the word. Then write the letters for the sounds you hear.

1. 

sk____t

3. 

f____n

5. 

p____se

2. 

st____

4. 

c____n

6. 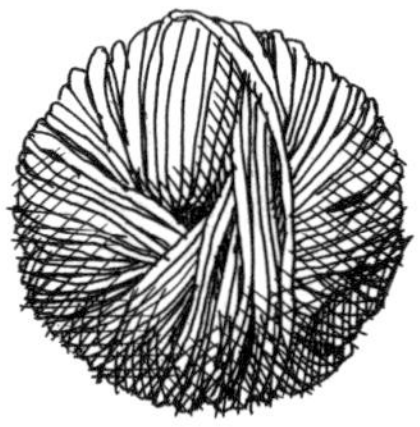

y____n

2. Circle all the words that are the same as the first one.

form	form	from	form	farm	form	force
chart	cart	charm	starch	chart	arch	start
herd	heard	hard	herd	her	herd	hers
sort	snort	sort	store	short	storm	sort
turn	turn	torn	runt	burn	turn	term
part	pert	park	part	tarp	port	part

3. Fill in the blanks with *er, ir, ur, ar,* or *or* to make a word.

1. ch _______ m
2. _______ ge
3. sn _______ t
4. j _______ k
5. b _______ p
6. c _______ n
7. y _______ d
8. tw _______ l
9. t _______ m

4. Use these words to fill in the blanks:

birth pork barn term
torn church sharp charge

1. Carl gives the clerk his ________ card to pay for the scarf.
2. Barb has not worn the skirt since it got ________.
3. For dinner, Mark had ________, corn, and tarts.
4. The farmer led his horse into a large ________.
5. A nurse helped Marge with her first baby's ________.
6. The Ford swerved on the ________ curve.
7. Arthur has started to serve his third _______ in office.
8. Fern parked her car not far from the ________.

Practice 50: ear and wor

1. **ear = /er/** **wor = /wer/**

heard	word
learn	work

2. Circle the word you hear.

1. word	cord
2. heard	hard
3. work	fork
4. yearn	yarn
5. earth	art
6. worm	form
7. worst	was
8. worth	north

3. Write the letters and say the word.

ear	wor
h ______ d	______ d
Ear l	______ k
p ______ l	______ m
______ n	______ se
l ______ n	______ st
y ______ n	______ th
s ______ ch	______ ld
______ th	

4. Read the words.

learn	worm
earn	worth
yearn	worse
pearl	worst
Earl	word
earth	world
search	work
heard	

5. Write the word you hear.

1. ______	9. ______
2. ______	10. ______
3. ______	11. ______
4. ______	12. ______
5. ______	13. ______
6. ______	14. ______
7. ______	15. ______
8. ______	

6. Read the sentences.
New words: worry, about

1. Earl works to earn money.
2. That string of pearls is worth a lot.
3. I yearn to visit the rest of the world.
4. Earl is learning to spell many words.
5. Ned is searching for worms for fishing.
6. Worrying about it just makes it worse.
7. The cops search for the missing pearls.
8. This car is the worst! It never works!
9. I have not heard a word he said.
10. Do not worry! The worst is over!
11. The worms are digging in the earth.
12. She earns about fifty dollars a day.
13. We heard about Earl. He is very sick.
    We worry that he will get worse.
14. We learned about the earth's crust.
15. This work is not worth doing.

Practice 51: Compound Words

1. Look at these compound words.

sunlamp	sun	lamp	hairbrush	hair	brush
fingerprint	finger	print	inchworm	inch	worm
slingshot	sling	shot	quicksand	quick	sand

2. Put the two words together to make a compound word. Read the word.

hand	stand	__________	land	mark	__________	sand	paper
work	bench	__________	egg	plant	__________	out	smart
left	over	__________	her	self	__________	sun	glasses
sports	cast	__________	back	track	__________	wind	storm

3. Find the two smaller words in each compound word. Write them in the blanks. Read the smaller words and the compound word.

riverbank	__________	__________	cornstarch	__________	__________
stopwatch	__________	__________	underwater	__________	__________
snapshot	__________	__________	pitchfork	__________	__________
drumstick	__________	__________	standstill	__________	__________
windchill	__________	__________	yardstick	__________	__________
grandfather	__________	__________	blacksmith	__________	__________

4. Make compound words. Match each word on the left with a word on the right. Write the compound word in the blank. The first one is done for you.

flash	flashback	stop	silk	silkworm	mother
witch	__________	mark	step	__________	worm
short	__________	car	play	__________	born
birth	__________	back	milk	__________	west
street	__________	grass	first	__________	man
crab	__________	craft	north	__________	thing

Practice 52: Two-Syllable Words

1. The big words below have been divided into syllables. Using what you have learned about sounding out words, sound out the syllables, and figure out each word.

plastic **/plas** tic/ skillet **/skil** lut/ clever **/clev** er/ intend /in **tend/**

2. Read the word first. Then read the sentence.

1. constant	**/con** stant/	We cannot stand that constant yelling!
2. product	**/prod** uct/	This shop sells a good product for less.
3. clinic	**/clin** ic/	This nurse works in a big clinic.
4. credit	**/cred** it/	Dad hands the clerk his credit card.
5. intend	/in **tend/**	Jan does not intend to marry Frank.
6. skillet	**/skil** lut/	Melt the butter in the hot skillet.
7. profit	**/prof** ut/	Dan made a profit on the stock market.

3. Read the word. Fill in the blank. Read the sentence.

1. pumpkin	**/pump** kin/	Mom is making ________________ bread.
2. current	**/cur** runt/	Yesterday's newspaper is not ________________.
3. selfish	**/self** ish/	Do not be ________________! Give it to me!
4. problem	**/prob** lum/	I tell my ________________ to my best friend.
5. suburb	**/sub** urb/	They live in a ________________ of a big city.
6. plastic	**/plas** tic/	The cups and dishes are ________________.

4. Read the story.

My Aunt Marge runs a factory in the suburbs.
Her factory makes pots, pans, and skillets.
She is working on her current product, a new skillet.
She says, "Ours will be the best product on the market!
We will run constant tests on it to get rid of any problems."
I have to give Aunt Marge a lot of credit.
She is smart and a hard worker.
She does not just make skillets.
She thinks up clever ads to sell them.
As you can guess, my aunt's factory makes a big profit!

5. Fill in each blank with one of these words.

frantic	**/fran** tic/	1. A cop is stopping the ________________.
sandwich	**/sand** wich/	2. The ________________ hid some nuts.
trumpet	**/trum** put/	3. Mother was ________________ with worry.
traffic	**/traf** fic/	4. Fran plays a ________________ in the band.
squirrel	**/squir** rul/	5. Put the ________________ in a plastic bag.
gravel	**/grav** ul/	6. The dump truck is carrying ________________.

Practice 53-A: CC-le Words with a Double Consonant in the Middle

1. The big words below have been divided into syllables. Using what you have learned about sounding out words, sound out the syllables and figure out the words.

apple **/ap** pul/ little **/lit** tul/ shuffle **/shuf** ful/ tickle **/tic** kul/

2. Read the word first. Then read the sentence.

1. saddle	**/sad** dul/	<u>Saddle</u> up the horses.
2. wiggle	**/wig** gul/	The baby <u>wiggles</u> his legs.
3. pickle	**/pic** kul/	Put some <u>pickles</u> on the bun.
4. kettle	**/ket** tul/	Mom is making jelly in a big <u>kettle</u>.
5. puddle	**/pud** dul/	John fell in the <u>puddle</u> and got wet.
6. tackle	**/tac** kul/	Ed takes his <u>tackle</u> box when he fishes.
7. nibble	**/nib** bul/	My son <u>nibbles</u> on the chips.

3. Read the word. Fill in the blank. Read the sentence.

1. bottle	**/bot** tul/	A ________________ of ink is on his desk.
2. tickle	**/tic** kul/	My baby laughs when I ________________ her.
3. rattle	**/rat** tul/	The baby plays with a ________________ .
4. puzzle	**/puz** zul/	Did you put the ________________ together?
5. struggle	**/strug** gul/	Dan had to ________________ to pass the test.
6. settle	**/set** tul/	Fred wants to ________________ in this city.

4. Read the story.

My little son Mark is three.
Yesterday I was tickling him.
He wiggled out of my arms and ran out into the yard.
Then he sat down in a mud puddle!
After lunch, I said, "Settle down with a puzzle."
"No!" Mark said. He did not want any puzzles.
I heard a rattle. Mark was playing with my pans and kettles.
I said "Stop!" and handed him an apple to nibble on.
Next I heard a crash! He had dropped a bottle of pickles.
Living with a boy of three can be a struggle!
But I want another kid just like him!

5. Fill in each blank with one of these *-le* words.

buckle	**/buc** kul/	1. Give me a pack of ________________ gum.
cattle	**/cat** tul/	2. I rest in the ________________ of the day.
raffle	**/raf** ful/	3. The river has lots of ________________s.
bubble	**/bub** bul/	4. Did Stan win a car in the ________________?
middle	**/mid** dul/	5. The belt ________________ is made of brass.
ripple	**/rip** pul/	6. Ted has a large herd of ________________.

Practice 53-B: CC-le Words with Two Different Consonants in the Middle

1. The big words below have been divided into syllables. Sound out the syllables and figure out the words.

handle /**han** dul/ gamble /**gam** bul/ single /**sing** gul/ ankle /**ang** kul/ whistle /**whis** ul/

2. Read the word first. Then read the sentence.

1. tangle	/**tang** gul/	The yarn is tangled up.
2. stumble	/**stum** bul/	Dad stumbled over the trash can.
3. sample	/**sam** pul/	I will sample the punch.
4. twinkle	/**twing** kul/	We look at the stars twinkle.
5. swindle	/**swin** dul/	He swindled her out of her money.
6. candle	/**can** dul/	It was dark, so I lit a candle.
7. gamble	/**gam** bul/	Mark never wins when he gambles.

3. Read the word. Fill in the blank. Read the sentence.

1. simple	/**sim** pul/	This math is ________________, not hard.
2. jungle	/**jung** gul/	The men walked out of the ________________ .
3. scramble	/**scram** bul/	________________ the eggs for dinner.
4. bundle	/**bun** dul/	Kim carries a ________________ under her arm.
5. angle	/**ang** gul/	Cut it on an ________________ .
6. tremble	/**trem** bul/	Her hands started to ________________.

4. Fill in each blank with one of the *-le* words.

handle	/**han** dul/	1. The king lives in a ________________ .
thimble	/**thim** bul/	2. They go to church in the ________________ .
ankle	/**ang** kul/	3. Pick it up by the ________________ .
castle	/**cas** ul/	4. Hank is not married. He is ________________ .
temple	/**tem** pul/	5. Chuck fell and turned his ________________ .
single	/**sing** gul/	6. The ________________ is big for my finger.
uncle	/**ung** kul/	7. ________________ Bob is Dad's brother.

5. *r*-C-*le* words. Read the word and the sentence.

1. turtle	/**tur** tul/	Barb gets a turtle at the pet shop.
2. marble	/**mar** bul/	The children are playing marbles.
3. girdle	/**gir** dul/	They sell girdles at the dress shop.
4. sparkle	/**spar** kul/	Your ring sparkles!
5. purple	/**pur** pul/	Her skirt is pink and purple.
6. startle	/**star** tul/	The telephone ring startled me.

Shopping at the Department Store

New words: Carter, apartment, department, store, handle

Marge Carter lives in a large apartment.
She says, "This is the day to go shopping.
I will go to the new department store I have heard about.
A person can get many things in a large department store."

Marge hurries out of the apartment.
She starts her car and heads for the department store.
The department store is on North First Street.
It is not far from the apartment.
She turns her car into the large parking lot and parks.

Marge gets a cart and starts shopping.
First Marge looks at a box of gift cards. It is marked $2.00.
She puts the cards in her cart.
Then she picks up some brushes.
"My son Carl will want some brushes for his art class," she says.

The third thing she picks up is a scarf.
The scarf is made of red yarn.
"My girl Barb will like this scarf," she says.

Then Marge looks at some shirts.
She thinks, "Mark wants a new sports shirt. His red one is torn."
She puts a shirt in the cart.

Next she picks out some kitchen curtains.
And she picks up some jars for canning.
Last she picks up some ferns and other plants for the apartment.

The clerk asks Marge, "Will this be cash or charge?"

Marge says, "I am short of cash."

She takes a charge card out of her purse and hands it to the clerk.
"It is smart of me to pay with my charge card."

"Yes," says the woman in back of her. "It can be smart to charge it.
But some cannot handle charge cards very well.
They charge many large bills.
They can pay for just part of their bills.
But they charge and charge until it gets hard for them to pay.
When they cannot pay, the store takes back their things.
Their charge cards have hurt them."

"Yes," Marge says. "You have to watch out when you charge things.
I am glad I can pay for what I charge.
Charge cards will never harm me.
I can handle them well."

Marge hurries out of the store with two large bags in her arms.
She thinks, "You have to be pretty smart just to shop in a department store!"